Immediate Response

MAJOR MARK HAMMOND DFC RM
as told to CLARE MACNAUGHTON

PENGUIN BOOKS

PENGUIN BOOKS

Published by the Penguin Group
Penguin Books Ltd, 80 Strand, London WC2R ORL, England
Penguin Group (USA) Inc., 375 Hudson Street, New York, New York 10014, USA
Penguin Group (Canada), 90 Eglinton Avenue East, Suite 700, Toronto, Ontario, Canada M4P 2Y3
(a division of Pearson Penguin Canada Inc.)
Penguin Ireland, 25 St Stephen's Green, Dublin 2, Ireland (a division of Penguin Books Ltd)
Penguin Group (Australia), 250 Camberwell Road, Camberwell, Victoria 3124, Australia
(a division of Pearson Australia Group Pty Ltd)
Penguin Books India Pvt Ltd, 11 Community Centre, Panchsheel Park, New Delhi – 110 017, India
Penguin Group (NZ), 67 Apollo Drive, Rosedale, North Shore 0632, New Zealand
(a division of Pearson New Zealand Ltd)
Penguin Books (South Africa) (Pty) Ltd, 24 Sturdee Avenue, Rosebank, Johannesburg 2196, South Africa

Penguin Books Ltd, Registered Offices: 80 Strand, London WC2R ORL, England

www.penguin.com

First published by Michael Joseph 2009
Published in Penguin Books 2010
2

Typeset by Palimpsest Book Production Limited, Grangemouth, Stirlingshire
Printed in England by Clays Ltd, St Ives plc

ISBN: 978-0-141-03904-6

www.greenpenguin.co.uk

For my wife and kids

Glossary

AFCS	Automatic Flight Control System
AH	Attack Helicopter
AI	Attitude Indicator
AMB	Air Mission Brief
ANA	Afghan National Army
ANSF	Afghan National Security Forces
APU	Auxiliary Power Unit
ASaC	Air Surveillance and Control
ASI	Airspeed Indicator
BARALT	Barometric Altimeter
C_2	Command and Control
CAP	Caution Advisory Panel
CAS	Close Air Support
Cherry	hot landing site, friendly forces in contact with enemy
CHF	Commando Helicopter Force
CINS	Chinook Integrated Navigation System
CP	Control Post
CTCRM	Commando Training Centre Royal Marines
Cyalumes	A company name, and the items can also be referred to as 'lightsticks', 'glowsticks' or similar
DC	District Centre
det	detachment
DFac	Dining Facility

EAPS	Engine Air Particle Separators
ECLs	Engine Control Levers (throttles)
EKIA	Enemy Killed In Action
EMR	Emergency Resupply
EOD	Explosive Ordnance Disposal
FOB	Forward Operating Base
FragO	Fragmentation Order sheet
GMLRS	Guided Multiple Launch Rocket System
Green Brain	aircrew information folder
Green Zone	vegetated area around the Helmand River
grots	accommodation
HALS	Hardened Aircraft Landing Strip
helo	helicopter
Hesco Bastion	tradename for blast protection walls – wire containers filled with rubble
HHI	Helicopter Handling Instructor
HLS	Helicopter Landing Site
hook master	electric switch controlling power to the hooks
HRF	Helmand Reaction Force
HSI	Horizontal Situation Indicator
Ice	cold landing site, safe to approach
IFF	Identification Friend or Foe
IFV	Infantry Fighting Vehicle
ILS	Instrument Landing System
Int	Intelligence
IRB	Infra-Red Beacon
IRT	Immediate Response Team
ISAF	International Security Assistance Force
JDAM	Joint Direct Attack Munition

JENGO	Junior Engineering Officer
JTAC	Joint Terminal Attack Controller
KAF	Kandahar Airforce Base
LCJ	Life-saving Combat Jacket
loc stat	location status
LZ	Landing Zone
MAOTs	Mobile Air Operations Team
MERT	Medical Emergency Response Team
MOD	Ministry of Defence
MOG	Mobile Outreach Group
MT	Motor Transport
NAS	Naval Air Squadron
NR	indicator of rotor speed
NSE	National Support Element
NVG	Night Vision Goggles
OC 'B'	Officer Commanding 'B' Flight
OC Forward	Officer Commanding JHF(A) Forward in Camp Bastion
OCF	Operational Conversion Flight
Ops room	Operations room
POC	Potential Officers Course
PONTI	Person Of No Tactical Importance
Prelim Ops	Preliminary Operations
PTIT	Power Turbine Inlet Temperature
PU	Power Unit
QHI	Qualified Helicopter Instructor
QHTI	Qualified Helicopter Tactics Instructor
QRF	Quick Reaction Force
RADALT	Radar Altimeter
RC(S)	Regional Command (South)
REMF	Rear-Echelon Mother Fucker
RIP	Relief in Place

ROC	Rehearsal of Concept
roulement	Deployment Rotation (French)
ROZ	Restricted Operating Zone
RTB	Return To Base
SA	Situational Awareness
sangar	a hardened defensive position, generally used by perimeter guard force
Sig Int	Signals Intelligence
SITREP	Situational Report
SOP	Standard Operating Procedure
T1	casualty requires immediate care for life-threatening injuries
T2	casualty requires urgent care for life-threatening injuries
TQ	Theatre Qualification
Ts and Ps	Temperatures and Pressures
UAV	Unmanned Aerial Vehicle
VNE	Velocity Never Exceed
VSI	Vertical Speed Indicator
WOG	Winch Operator's Grip

Boeing Vertol
Chinook HC Mk1

1 FM homing aerials
2 Pitot tubes
3 Nose compartment access hatch
4 Vibration absorber
5 IFF aerial
6 Windscreen panels
7 Windscreen wipers
8 Instrument panel glare shield
9 Rudder pedals
10 Yaw sensing ports (automatic flight control system)
11 Downward vision window
12 Pilot's footboards
13 Collective pitch control column
14 Cyclic pitch control column
15 Co-pilot's seat
16 Centre instrument console
17 Pilot's seat
18 Glideslope aerial
19 Forward transmission housing fairing
20 Cockpit overhead window
21 Doorway from main cabin
22 Cockpit emergency exit doors
23 Sliding side window panel
24 Cockpit bulkhead
25 Vibration absorber

38 Blade drag dampers
39 Glassfibre rotor blades
40 Titanium leading edge capping with de-icing provision
41 Rescue hoist/winch
42 Forward transmission aft fairing
43 Hydraulic system modules
44 Control levers
45 Front fuselage frame and stringer construction

53 Cabin window panel
54 Cabin heater duct outlet
55 Troop seats stowed against cabin wall
56 Cabin roof synchronising shaft
57 Formation keeping lights
58 Rotor blade cross-section
59 Blade balance and tracking weight socket
60 Leading edge anti-erosion strip
61 Fixed tab
62 Fuselage skin plating
63 Maintenance walkway
64 Transmission tunnel access doors
65 VHF/AM – UHF/AM aerial
66 Troop seating, up to 44 troops
67 Cargo hook access hatch
68 VOR aerial
69 Cabin lining panels
70 Control runs
71 Main transmission shaft
72 Shaft couplings

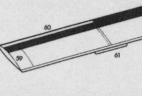

26 Cockpit door release handle
27 Radio and electronics racks
28 Sloping bulkhead
29 Stick boost actuators
30 Stability augmentation system actuators
31 Forward transmission mounting structure
32 Windscreen washer reservoir
33 Rotor control hydraulic jack
34 Forward transmission gearbox
35 Rotor head fairing
36 Forward rotor head mechanism
37 Pitch change control levers

46 Emergency exit window
47 Forward end of cargo floor
48 Fuel tank fuselage side fairing
49 Battery
50 Electrical system equipment bay
51 HF/SSB aerial cable
52 Stretcher rack (up to 24 stretchers)

73 Centre fuselage construction
74 Centre aisle seating (optional)
75 Main cargo floor: 1,440 cubic ft (40.78m³) cargo volume
76 Ramp-down 'dam' for water-borne operations
77 Ramp hydraulic jack

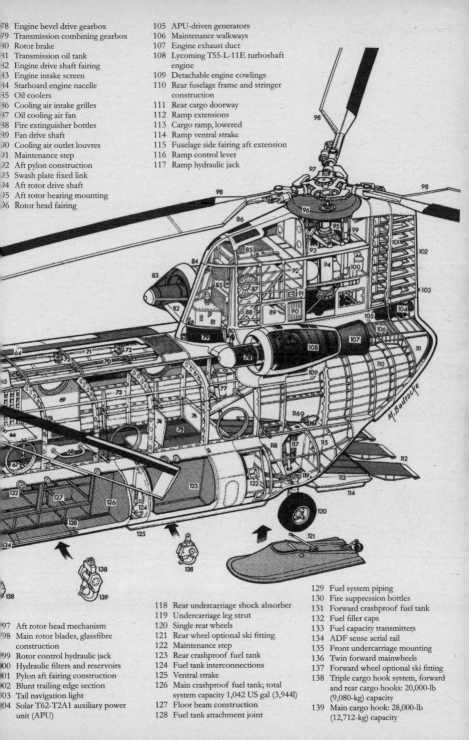

78 Engine bevel drive gearbox
79 Transmission combining gearbox
80 Rotor brake
81 Transmission oil tank
82 Engine drive shaft fairing
83 Engine intake screen
84 Starboard engine nacelle
85 Oil coolers
86 Cooling air intake grilles
87 Oil cooling air fan
88 Fire extinguisher bottles
89 Fan drive shaft
90 Cooling air outlet louvres
91 Maintenance step
92 Aft pylon construction
93 Swash plate fixed link
94 Aft rotor drive shaft
95 Aft rotor bearing mounting
96 Rotor head fairing

105 APU-driven generators
106 Maintenance walkways
107 Engine exhaust duct
108 Lycoming T55-L-11E turboshaft engine
109 Detachable engine cowlings
110 Rear fuselage frame and stringer construction
111 Rear cargo doorway
112 Ramp extensions
113 Cargo ramp, lowered
114 Ramp ventral strake
115 Fuselage side fairing aft extension
116 Ramp control lever
117 Ramp hydraulic jack

M. Badrocke

97 Aft rotor head mechanism
98 Main rotor blades, glassfibre construction
99 Rotor control hydraulic jack
100 Hydraulic filters and reservoirs
101 Pylon aft fairing construction
102 Blunt trailing edge section
103 Tail navigation light
104 Solar T62-T2A1 auxiliary power unit (APU)

118 Rear undercarriage shock absorber
119 Undercarriage leg strut
120 Single rear wheels
121 Rear wheel optional ski fitting
122 Maintenance step
123 Rear crashproof fuel tank
124 Fuel tank interconnections
125 Ventral strake
126 Main crashproof fuel tank; total system capacity 1,042 US gal (3,944l)
127 Floor beam construction
128 Fuel tank attachment joint

129 Fuel system piping
130 Fire suppression bottles
131 Forward crashproof fuel tank
132 Fuel filler caps
133 Fuel capacity transmitters
134 ADF sense aerial rail
135 Front undercarriage mounting
136 Twin forward mainwheels
137 Forward wheel optional ski fitting
138 Triple cargo hook system, forward and rear cargo hooks: 20,000-lb (9,080-kg) capacity
139 Main cargo hook: 28,000-lb (12,712-kg) capacity

Part 1

Prologue

6 September 2006

The voice of the Control Post (CP) watchkeeper cut across the airwaves.

'Luke Skywalker . . . this is Deathstar . . . Move to our location now.'

I turned round. 'Gents, we've got a shout.'

I glanced at my watch. 1700. Looked as if we'd be missing scran tonight. The room was silent as we focused on gathering our shit together and getting quickly over to the CP.

I pulled on my desert boots. My Browning 9mm automatic pistol with two mags was hanging on the front of my cot and above it was a clothes line with my shirt, sweat rag and dog tags. I grabbed the holster and strapped it around my thigh. Then I shoved in the handgun and secured the plastic release clip. As I got ready I was excited and nervous. I could feel the adrenaline begin to surge through my body. Five minutes later we were ready.

'Right,' I said, 'let's go. I've got the radio.'

We walked out of the tent and over to a metal ladder, chopped in half and resting against the Hesco Bastion wall, a 5ft-high wall made from wire-framed mesh containers, lined with a very tough canvas-type material, filled with rubble. The wall was designed to stop shrapnel, explosives and other devices intended to kill us and was a constant reminder of the threat we were surrounded by. A genius bit of kit that has made its British inventor, quite rightly, very rich. I climbed on to the top of the wall and jumped over to the road.

Dan leapt into the Land Rover Defender soft-top Safari and moved it so it was parked right outside the entrance to the CP. We were now lined up with a vehicle ready to get to the cab – the helicopter – ASAP.

Daffers and I ran as fast as we could towards the CP so I could grab the Green Brain and my pilot's kneeboard from next to the bird table – which supported the detailed, regularly updated map of the operational area showing contacts and the disposition of all forces. The Green Brain contains location call signs, Fragmentation Order sheet (or FragO), the tail number of your aircraft, Identification Friend or Foe codes and mission numbers for you and for all the other aircraft operating out of Bastion. We simply couldn't operate without it.

'Where's Mike?' I asked the Corporal as we left the CP. Major Mike McGinty was OC Forward. He was the Apache gunship Squadron Commander and ran the Forward area. Average height, slim build, black hair, good operator, highly motivated with a great sense of humour. I had known him for years.

'He's next door already.'

We legged it into the 3 Para CP. It was a hive of activity. Everyone had closed up to the bird table to get an update on what was going on.

The CO was Lt Col. Stuart Tootal, Commanding Officer of 3 Para Battle Group, a short fella with grey hair. He is a good officer and a great leader. Tootal had been in theatre for four months and had already lost nine men from his unit. The strain was beginning to show on his face.

I walked into the room. The Regimental Sergeant Major was there. He was a bear of a man with a handlebar moustache. Neither the Sergeant Major nor Tootal acknowledged us. We were the new kids on the block. And we wanted to leave a good impression and show them we were consummate professionals. They had been out there for a while and for us this was our first step up to the plate. It was important not to come across as either inexperienced tossers or, worse, clowns who thought they knew it all.

I hung back. Tootal was trying to get a report from the medical guys in Sangin and McGinty was up at the table.

I whispered over to one of the watchkeepers, a young guy with black hair: 'What have we got?'

He glanced over to me and replied, his voice hushed: 'An RPG has been fired into one of the sentry sangars at Sangin' – in other words, a rocket-propelled grenade fired at a protected sentry position. 'We have one T1 and two T2s.'

Instantly I thought about the threat. I wanted to hear what the Intelligence guys had to say about what was going on. As it was a T1 I knew that we would be going for sure. A T1 is a casualty who will die if we cannot get them to immediate life-saving medical treatment; so I waited to hear from the Apaches which direction we would be going in from.

Within minutes I had the Intelligence update.

And I knew then it was going to get sporty. The Taliban would be armed and waiting for us. It would be a huge morale boost to them to bring down a Chinook so I knew they weren't just going to let us cruise in, collect our guys and then rock on out of there. We would definitely be feeling the full force of their firepower.

Hoofing, I thought.

Daffers and I exchanged a look that said: 'Hmm, this is going to be interesting.'

I went forward towards the table to speak with Mike McGinty about how we were going to go in. Tootal looked on and occasionally interjected.

'What do you want to do?' he asked. 'Which direction do you want to come in from?'

I thought about it but not for long. I didn't want to show too much hesitation. 'We will go out high, drop down and then come in from the west.'

This was mainly because that was the way that I had gone into Sangin before so I knew I could get in quick with the minimum of navigation. There was barren desert up on the west and it was the safest transit into the District Centre (DC). Any other direction would take us into the Green Zone. We would want to minimize our time there because that is where the Taliban are based and we were more likely to get shot at.

'We'll be on station two minutes ahead,' McGinty added.

That was crucial. Without the Apaches providing top cover we'd be going nowhere.

I looked up at him. 'The comms will be yours and I need a cleared in and a hot or cold LZ call.' I needed to ensure that the Apaches talked to the guys on the deck to clear us into the restricted operating zone and make sure we had clear access to the Landing Zone.

I turned to the watchkeeper. 'Are the medics at the cab?'

He nodded. I turned to the Colonel. 'You happy, Boss?'

'Yeah, good to go,' he confirmed, with a look that said: '*Yes, just get on with it and get my boys back.*'

Daffers and I turned and made our way outside. As we broke into a run, I checked that we were both in sync: 'And you're happy with the plan?'

He nodded.

I continued: 'I'll go in the right, you go in the left. Have you got a grid for Sangin?'

'It should be in the kit?'

'We need to check that,' I said as we got to the Land Rover. Sam was waiting behind the wheel looking expectant. He wanted to know what was going down. I jumped into the left-hand seat, Dan and Daffers into the back. They were barely sitting before Sambo put his foot down. The Landie was chucking up dust as we tear-arsed towards the cab. Around camp, foot down meant about 50 mph and as the speed limit is 10 mph we were going some. But everyone knew we were the Immediate Response Team (IRT) vehicle. Normal rules didn't apply. I turned to the crew and gave Sam and Dan a heads-up that we had a T1 at Sangin.

Two minutes later we arrived at the cab – the big, dark-green twin-rotor Boeing Vertol CH-47 Chinook helicopter. The engineers were swarming all over, checking it and making sure that everything was ready.

Sam parked at the rear of the helo and I jumped down. I did a quick walk round the aircraft to check for any obvious defects, leaks and faults, then signed the book to take responsibility for her and jogged up the Chinook's tail ramp grabbing my kit as I passed it, hanging up from a seat rail inside the big cargo hold.

The medical team of five and our force protection team (the guys from 3 Para who are there to provide additional fire support when we're on the ground loading the casualties) all followed me in and grabbed seats along the sides of the cab. I moved forward into the cockpit as I pulled on my LCJ, a green survival vest with a 'bullet bouncer', or body armour, in it. I hauled myself into the cockpit, squeezed myself into the right-hand armoured seat, put in my ceramic chest plate, zipped it up and pulled my five-point harness together, slotting the lugs into the harness box between my legs. Before bringing the big Chinook to life I glanced back and shouted a warning: 'Helmets!' Then I turned on the battery. Immediately the RADALT – radar altimeter – audio warning started warbling in our ears. I reached down automatically and pressed the button on the collective lever to cancel it. Daffers and I raced through the pre-flight checklist, focused on being airborne in a heartbeat.

Oh my God! I thought, *we are really doing this*, and once again I felt the adrenaline surge through my veins.

Chapter 1

28 August 2006

'Are you going back to Afghanistan?' asked my eldest daughter.

'Yes, I am,' I replied.

'Excellent, that means we can watch *America's Next Top Model!* – Yesss!' She raised her fist and pulled her arm quickly into her stomach as if her team had just scored a goal.

My youngest looked at me solemnly. 'And I can watch *Spirit?*' – her favourite DVD of the moment, about a wild stallion.

We had let the kids know that I was going away but it really was the norm to them. They were not that fussed as they were genuinely too wrapped up in their own world to be affected by mine. They knew that Daddy was going away for two months. Afghanistan was just a word to the kids. It's not a place. We didn't give them the details of what it was like.

'It's nice to see that I will be sorely missed by my children,' I said.

Not to mention that my television would be hosting some fine viewing. Undoubtedly my wife, aka Domestic Chief of Staff (DCOS), would be looking forward to watching *Big Brother* without a barrage of sarcasm from me. I genuinely believed the programme to be utter horse shit and I was incapable of keeping this to myself.

'How can you watch this?'

'Mark – shut up!'

'No, seriously – how can you watch this. It's awful!'

'Mark – shut up!' she snapped more curtly this time.

'They are a bunch of gormless, spotty oiks – it truly is dreadful.' I wouldn't let it go.

'Maaark! – I am trying to watch it!'

'Sod this, I am going down the gym.'

Ah. The harmony of family life.

I had been married for around eight years. About ten days before we left, the DCOS went into remote mode. This was when we began to distance ourselves from each other because she knew she was going to be on her own for two months. We called it being an 'inny' or an 'outy'. If you were an 'inny' then you were in the family and if you were an 'outy' then you were outside the circle of trust. About this time she kicked me out of the circle of trust and I became an 'outy'.

The DCOS was quiet and not really speaking to me. Out of the kitchen window I could see the sun shining brightly over the rolling fields on a fine August bank holiday weekend. The kids were chattering like monkeys, shouting, screeching and bouncing on the trampoline in the garden. It was noise pollution and I couldn't bear it.

I poked my head round the corner and yelled at them: 'Keep it down!'

They stopped, looked at me for about ten seconds and then carried on with the volume set at exactly the same level.

Yep, total control, I thought to myself and wandered off into the shit heap of a garage to drag out my Bergen and black box. As I stepped over the cartons of boxes I looked thoughtfully at the true love of my life, my Harley-Davidson – 'The Hog'. Barely used since the kids were born, a huge bone of contention between me and the DCOS, but nevertheless as much a part of me as my right arm. I loved that bike. No matter how much she nagged me I was never going to sell it. I took a moment to savour its beauty and then cracked on.

Back inside, I gathered the stuff which was most important to me – comfy flip-flops, a head torch, my American poncho to adorn my bed, providing a splash of colour and at the same time additional warmth, an extra pillow, some grot mags and my iPod and speakers. These little comforts were the things that were going to make my life in theatre a bit more pleasant.

It took time to pack. The DCOS didn't have anything to do with it, apart from the odd argument about where my kit was. I had washed all my desert combats, flying T-shirts, socks, pants and flying suits.

'Have you seen my beret?' I asked.

'No!' she replied abruptly.

'All right, everyone's had a coffee. Keep your hair on. I only asked,' I said, in a slightly falsetto voice.

I was always losing my beret. Clearly it's never my fault. I am pretty sure that DCOS hides it on purpose so I can't find it.

'It's all right, I've found it,' I shouted to her.

It was on the coat hooks in the hall.

'Where was it?' she yelled back.

'It was underneath the kids' coats that you left on the stairs,' I fibbed. It wasn't but I wasn't going to let her know that.

I am much tidier in my military life than I am in my home life. In the Bergen everything had its place. Pants, socks, uniform were ironed ready to go. I laid them out around the living room until the DCOS couldn't take it any longer and freaked out at me.

'Jeez, Mark! Do we all have to share in this? Can't you pack it up quicker? Your shit is everywhere. We live here too you know,' she bitched at me in her dulcet American drawl.

We met while I was flying Cobra helicopter gunships on an exchange with the US Marine Corps in San Diego.

'All right, all right. I don't want to forget anything,' I said.

She threw me a murderous look and then walked into the kitchen.

I picked up the pace on the packing and then moved my bags, black box and Bergen into the passageway down the side of the house ready for the off.

As it was the bank holiday weekend the DCOS drove me up to the squadron. I'd hugged the kids goodbye but to them it was no different a goodbye from when they were going to school. It wasn't overly emotional for them so it wasn't overly emotional for me.

'Bye, sweetheart,' I said giving my eldest a hug.

'Byeeee,' she said and ran off.

'Bye, baby,' I said to my youngest. 'Be good for Mummy.'

'I will,' she said, looking at me very intently. 'I want to ask you something, Daddy.'

'Yes, baby?' I said.

'Can I be Spirit?' she said earnestly.

'Of course, sweetheart,' I said, laughing.

And with a quick flick of her mane, she raised up her hooves and trotted off. I tried not to think about how much I was going to miss my girls.

'See ya, fella,' I said to Flash, my German shepherd dog. He cocked one ear, looked at me and walked back into the garden. He would probably miss me more than my family would. Us boys have to stick together.

The DCOS and I didn't say much to each other on the short drive over to the squadron. We had been separated so many times before. I had done at least three six-month deployments since we had been married. These little six-week stints were nothing in comparison. The communications package is ok, there are phones and Internet, so if I could be arsed to queue up for ages then we could speak to each other. Generally, I phoned once a week. We were not big on that loads of communication thing. Sometimes she bent my ear about it but I rode it out and it passed.

Normally we would go out to dinner a few evenings beforehand and everything would be squared away in a civilized, relaxed atmosphere. On the day of departure there was very little more to be said.

We pulled up at the squadron. A few of the lads were hanging around. The DCOS looked at me from the driver's seat.

'Fly safe,' she said.

'I will.'

'I love you,' she said.

'I love you too,' was my reply.

I leant over, gave her a brief kiss on the lips and then got out of the car, pushing the door shut behind me. I opened the boot, dragged out my kit and slammed the lid shut. I banged the top of the roof. She gave me a wave and then drove off.

I walked over to the guys that were hanging around and reached inside my pocket to find my packet of cigarettes. As I lit my first Marlboro Red, I sighed to myself: 'Fucking hoofing.'

A fag and det Tourette's (the inability to speak without swearing, which only occurred when ensconced in a completely military environment). I knew the detachment had officially begun.

Sam Hannant, one of the crewmen, came over.

'Morning, mate, any chance you've got a fag I can nick?' He playfully punched me on the shoulder. I reached into my pocket and pulled out my Reds.

'Do you think you are man enough?' I responded, laughing and holding out the packet towards him.

Sambo was a great guy, kind of cuddly, with freckles, and a little bit ginger. He was my smoking buddy on det. He had a very dry sense of humour. He often said the funniest things at the most inappropriate of moments. He was a good lad to be spending six weeks at the Afghan beach with. He had been stood with Dan Baxter, the other crewman on my crew. Dan was an equally fine fellow, of middle height with dark hair. He was another gym lover. We had already agreed that we would be hitting the gym together.

'You all right, mate, you look threaders?' I said.

'Got dumped last night!'

'No way,' I said.

'I'm looking forward to some away time. Y'know, to keep my mind off it.'

'Don't worry, plenty more crocodiles in the sewer. We'll work it out in the gym. Give the guns some sickness,' I said trying to cheer him up.

My old pal Daffers strolled over. We had been paired up for the start of the det. He was to be my co-pilot. Daffers and I went way back. He was ex-Army Air Corps, had come up through the ranks and was subsequently given a commission in the RAF. There was an affinity between us because of his 'greener' Army background. We'd spent some time together at Middle Wallop so we had been here before many times.

'Hello, big fella. How's it going?' he said, grinning and reaching out to shake my hand.

'Same crap, different day,' I said.

'Can you believe it? When do we get August bank holiday weekends with weather like this?' he said incredulously.

'I should be sitting in my garden being wound up by children, nursing a cold beer,' I said.

We were all whingeing about the travel arrangements – no matter what service you were in, Navy, Army or RAF, everyone hated movers – as the white, standard, nondescript MT – motor transport – coach pulled up around the corner. 'Ah, our chariot has arrived,' said Daffers and wandered off to get his kit. We all got our bags together and started chucking them onboard the bus. Around thirty of us from 18 Squadron 'B' Flight were headed out. Everyone jostled about getting their boxes and Bergens into the luggage hatches. Eventually I climbed up the steps and snagged a seat to myself. I plugged the headphones into my iPod, put them on and stared out of the window.

RAF Odiham was in full bloom. The base was quiet and we didn't see a soul other than the guards as we drove out of the gate and up the road. An airfield had been on the site, nestled in the Hampshire countryside, since 1925. It was opened as an RAF Station in 1937 by General Erhard Milch, Chief of Staff of the German Luftwaffe. He was so impressed with what he saw that he is reputed to have told Hitler: 'When we conquer England, Odiham will be my Air Headquarters', and he ordered his pilots not to bomb it. Chance would be a fine thing, cheeky Jerry bastard!

The hardest part of saying goodbye to my family came when we were already separated. We turned out of the station. I watched my house disappear over the field as we drove to Brize Norton. It was then the sadness hit me and I knew that I wasn't going to be around the chaos and the warmth of family life for a few months. I didn't dwell on it though. This was the job. This was what came with it and you man up.

Chapter 2

As the RAF Tristar climbed away from Brize I turned to Daffers. 'It's weird. I am looking forward to getting on with it.'

'It's good to get out into theatre and lose yourself entirely for a while,' he agreed. 'Life is simpler. Eat, sleep, work, eat, sleep, work, sometimes just work, work, work.'

'Don't forget work out, work out, work out, chain smoke and work out,' I quipped. 'I have heard it's all changed since Prelim Ops, mate.'

We had been out on Prelim Ops in January 2006 and laid the foundations of the war machine we were now returning to. It had been more of a milk run, tasking here and there and working mainly out of Kandahar airbase.

Strangely, I was excited at the prospect of going back in. I don't wish harm on anyone but at the same time being in an operational theatre is an opportunity to exercise all your training in real-life situations, test your mettle and put your fighting pants on. It's kind of like surgeons: they don't want anyone to get ill and have bits cut out of them, but when someone does require surgery it's a great opportunity to put all that learning into practice.

The word on the street was that the contacts had got a lot hotter since then. 'I have always wanted to be shot at,' said Daffers. ' It's why I joined up.' He laughed laconically.

'Mate, you joined up cos no other idiot would employ you,' I said.

'Yeah, right. And you joined up because you don't need to

spell to be a Bootneck. The expression's not "sweating like a Bootneck in a spelling test" for nothing.'

'I can't help it if I am dyslexic! You shouldn't mock people with learning difficulties.'

We both laughed.

Bootneck is slang for Royal Marine and I was a Royal Marine on exchange with the RAF.

As I sat back to endure the 7½-hour flight that lay ahead of me I thought about the path that had led me there. I had known since the age of eleven that I wanted to join the Corps (that's the Royal Marine Corps to you). The Royal Marines are separate to the rest of the military. We are unrivalled and we wear the coveted green beret, an emblem of our elite-ness. Unlike many of the Army brigades and regiments the Royal Marines are classless: no trust funds or nepotism will get you into this club. It is a measure of the man alone to pass the Commando course.

At school, I was interviewed by the RM schools liaison officer. The recruiting officer said to me: 'Work hard, get your O-levels and we'll put you forward for an A-level schol-arship at sixteen.'

I said: 'No problem.'

He then gave me a basic interview, which looked at my overall suitability. I passed this. At fifteen I attended a Poten-tial Officers visit at 42 Commando in Plymouth. This week with an RM unit firmly cemented my burning desire to join the Corps.

At sixteen I went on to the Potential Officers Course (POC) at the Commando Training Centre Royal Marines (CTCRM), Lympstone. This was a three-day familiarization to see if I was physically and mentally up to the gig – a series of physical and leadership tests, fitness tests, presentations and this hoofing thing called Hare and Hounds, where the

Batch Officer ran ahead and we all had to catch him and stay with him for the duration. One thing about being an RM was that you never dropped back in training. Drop back and you were out.

I was ecstatic when I passed this, but then I was forwarded to the Admiralty Interview Board, and failed. I was gutted but I didn't give up. I wanted to be a Royal Marine more than anything I had ever wanted and this just made me want it even more. I was determined to serve in one of Britain's most elite military forces and I was determined to become one of the best-trained soldiers in the world.

At twenty-one, having completed the whole selection process yet again, I was finally offered my much-sought-after place at CTCRM. I started fifteen months of rigorous, ball-breaking training to join the biggest boys' club in the world.

The Royal Marines were such a small, tightly knit unit that it engendered great loyalty among its soldiers. There was something special and unique about being a Royal Marine and there was incredible kudos to being a Royal Marine officer. I entered the course a boy and came out a man.

Being 'Jack' – being out for yourself, not pulling your weight – was the biggest insult in the Royal Marines. As a result, throughout the entire training and work process we always strived to work as a team. This was a fundamental part of our ethos.

The outcome is that you become an officer and a leader of men. During training, officers did everything that a recruit would do but accelerated and more. It was expected that officers were better so we were trained harder.

The Commando course stretched me beyond the realms of the impossible. There were times when I thought that I wasn't going to make it. The pressure of the course and maintaining the standards they expected of me was unbearable. Like when I was still in the field at 2 a.m., sleep-deprived, sick with exhaustion, covered from head to toe in shit. I would think: *I don't know how I am going to get myself back to my cabin and get all my kit paraded in time*, knowing that in six hours, by 8 a.m., everything needed to be pristine. Nothing could be out of place and if they found so much as hair in your comb you would be fucked!

In training there was a Bootneck mantra impressed upon us: 'It pays to be a winner.' It was used as motivational punishment. Everyone raced up the road to a village and the first one back didn't have to go again. But unless you were the fastest in the group you needed to work out how many times you were going have to do the run. You didn't want to run too fast on any of the runs so that you were fucked up for the next one. This was a standard punishment and if you were the slowest you'd get beasted more.

Yet somehow I did it and they got me to a standard where I did have all my kit squared away in time ready for parade against all odds. The training team instilled in us the level the bar was set at. It was the standard to attain and if we didn't meet it then we were out. The only way to get through was to think: *Bring it on, you are not getting rid of me. I want that green lid more then I want to quit.* The green beret meant brotherhood. It was a badge of honour and being one of the Royal Marines brethren was my ultimate goal.

Royal Marine officers knew that they were going to command Infantry. You were going to kill people, ambush people and attack places so the troops on the ground had to know who was in charge. They had to know who the leader was, and as a Royal Marine Commando Troop Officer that person was you. This created an uncompromising mentality in which defeat was not an option. Royal Marine officers were very single-minded and they stopped at nothing to achieve an objective.

It was fundamental to your role to be a physically fit bloke. If you were stronger, you would be able to lead physically, which was a must for a Commando officer. If you were fit and robust, undoubtedly, you would be harder and more enduring than your troops. If the lads were feeling knackered you could say: '*If I can do it then it isn't that bad so man up, wet pants!*'

With my parents watching, and in the proudest moment of my life, as a Second Lieutenant, having successfully completed the Commando Course, I was awarded my coveted green beret and passed out from CTCRM Lympstone in 1990. I was then posted as Officer Commanding 4 Troop, Lima Company, 42 Commando RM, in charge of thirty hard-charging, glass-chewing, hairy-arsed Bootnecks.

Flying came later and I was incarcerated in a desk job for eighteen months until they allowed me to start the course. It was no secret in the military how much I loathed staff work. All I ever wanted to do was be a Royal Marine and then a pilot. There was nothing that pissed me off more than staff work and desk jobs.

There were around sixty serving Royal Marine helicopter pilots. Royal Marine aviators formed part of the Fleet Air Arm. They predominantly served as pilots within the Commando Helicopter Force (CHF). This was based at RNAS Yeovilton in Somerset and was made up of three Sea King Naval Air Squadrons (NAS), 845, 846 and 848, as well as 3 Commando Brigade Air Squadron (3 BAS), which was recommissioned as 847 NAS. Manned by Royal Marines and flying borrowed Army Air Corps aircraft we were the red-headed, banjo-playing stepchildren of the military aviation world.

Since then and as I hated staff work, I didn't want to follow the typical RM aviation career, which inevitably would imprison me behind a desk. I wanted to stay in the cockpit. I had flown attack helicopters with the US Marine Corps. I had flown utility helicopters, the Lynx and Gazelles, with 3 BAS and 847 NAS. I'd had a lot of good flying. But it was only going to get harder to keep steering clear of that desk.

Then one of my mentors and a thoroughly good egg, a former boss and great mate of mine called Bill O'Donnell, suggested that I fly the Chinook. At that time I was thinking of throwing in the towel, leaving the Corps and heading to the sunny shores of Australia. Bill dissuaded me and said that my talents could be used on the CH-47 exchange, which had been running for ten years. Three extra billets had been arranged on it because there was talk of CHF purchasing some CH-47s and it wanted naval/RM aviators to gain some knowledge of the CH-47 in case the plan came to fruition. I was fortunate to be offered a place on the exchange, which I snapped up.

Chapter 3

'We will shortly be arriving in Kandahar,' a voice announced over the Tristar loudspeaker. 'It's time to put your body armour and helmets on.'

I stood up and pulled my body armour over my shoulders, put my helmet on and fastened the strap under my chin. I sat down and fastened my seat belt. The lights went out and we began our descent.

The news on the grapevine was that things were hotting up and I wasn't just talking about the blistering heat of the sun. It was no longer a milk run. Over the summer we heard from Woodsy's flight ('A' Flight, 18 Squadron) and the guys on the ground that it was becoming increasingly dangerous.

Our troops on the ground were entrenched in platoon houses in the districts of Helmand: Musa Qala, Now Zad and Kajaki in the north, Sangin to the south-west and Garmsir in the south. The platoon houses acted as bullet magnets to the Taliban, who consistently suffered heavy losses trying to take them. For 3 Para, the battle group currently on the ground, it meant they were holed up day after day, week after week, month after month, in bandit country having the crap bombarded out of them. Withdrawing from the platoon houses would have been a symbolic victory for the Taliban, so our job was to get the guys the ammunition and supplies they needed and also to evacuate any casualties as quickly as possible so that they could continue to hold their position. Sangin was bad but Musa Qala was worse – it was like the Alamo of Afghanistan. The troops were oiling weapons with vegetable oil because they had no supplies. Their only life-line was the Chinooks.

The Taliban fire was so heavy and accurate in these areas that the cabs faced a serious risk of being hit. By mounting air supply and casualty evacuation missions there was a real possibility of losing an aircraft. We had very reliable Intelligence that informed us that the Chinook was a major target for the Taliban and for them to bring one down would be a considered a significant victory in their eyes. With only eight cabs in theatre this would also be very bad for us on many levels.

Kandahar was our main operating base, from where we would primarily transport regular resupply runs or assist with the movement of troops in and out of the districts. There was a Forward Operating Base (FOB) at Camp Bastion in Helmand Province and from here the role of the Chinook was to operate as the Immediate Response Team (IRT) and Quick Reaction Force (QRF) and be on 24-hour standby to respond to any casualty evacuations or any other immediate battlefield requirements. We would be splitting our time between Kandahar and Bastion.

The Teletubbies – the Taliban – were not idiots and the Chinook is not a subtle aircraft. It's around 100ft long and maxes out at about 180 mph, so it is a big, cumbersome, relatively easy target. As an aircraft it's incredibly well defended, with a huge amount of system redundancy, lots of armour and arguably the best self-defence suite of any aircraft in the military. It's designed to soak up bullets and still keep flying. I'd never flown anything like it before.

The Boeing CH-47 Chinook is a helicopter that always gets a reaction from people when they first see it. Some people think it is ugly and dysfunctional. Others, like me, see the beauty in its function – from the six blades, where the front three intermesh with the rear three, all designed to withstand having large chunks shot off them, to the two large engines stuck to the side of the fuselage like pods so they can be on fire without damaging the rest of the aircraft. It is a cab built for combat and we were going to be doing a lot of that.

I seized the opportunity to fly the Chinook. There were not many exchanges in the military like this one and I was one lucky Royal Marine to be able get my hands on this aircraft because they are flown only by the RAF. They are awesome beasts to fly and it is the only aviation unit in the UK armed forces that has been on continuous operations, without a break, for twenty-six years.

Chinooks have been active in a huge variety of countries, roles and operations. They have operated covertly, they have recovered hostages, they have provided humanitarian support in Pakistan and Turkey, flown in the jungle of Sierra Leone, evacuated Lebanon, and fought in both Gulf Wars and in the Cold War. The variety of operations participated in by the Chinooks is spectacular and unparalleled by any other force of any type in the Royal Air Force. Within the Chinook fleet there are wizened aviators who have been flying with the Chinooks since 1981. They have chests full of medals, because they have been everywhere and done everything and what they bring to the party is advice to the commanders, pilots and captains. Regardless of their rank they are the wise men who advise us on what works and what doesn't. It's essential that the Chinook force is at the top of its game.

The CH-47's are fundamental to the entire Afghanistan campaign. It is one of two or three key battle-winning capabilities. We are delivering vital food, water, ammunition and fuel to remote places. Without this delivery service the campaign can't carry on. The Shuras, or councils, can't go on, the elders can't meet, the governor can't arrive safely to talk to the local population. The whole development of Afghanistan comes to a halt if we can't get Chinooks on the ground at the time and place of our choosing.

Even though we are also involved beyond the battle, within the battle we are key. There are many senior Army officers who have said: 'But for the Chinook force we would have died.'

Of some 2000 casualties lifted, without the Chinooks 300 would be dead. If the news of 300 more deaths in the campaign were broadcast by the newspapers back home, the population of the UK would have a very different perception of the conflict. Afghanistan as a valuable and valid operation would then be drawn further into question. This emphasizes the strategic value of the Chinook in the campaign, not just its tactical worth, and how crucial the aircraft, the crews and the people who support them are.

Among the Taliban, we can see combat Darwinism in action. The clever ones are the ones that are left alive. They are getting smarter. The more they observe us the more they understand how we operate and think. They know we have legal constraints and they can actively exploit the weaknesses created by those constraints. The Taliban are targeting the Chinooks and we are the ones taking the rounds.

'Cabin crew – ten minutes to landing,' said the Captain over the intercom.

The aircraft screeched a little as its wheels bumped down on to the tarmac and we were thrown around in our seats.

I turned to Daffers. 'At least no one took a pop at us on the way in.'

Chapter 4

The heat hit us as we stepped off the Tristar. After we got our bags we were met by Digs, a tall gangly fellow, and a very British officer, who was the boss of 27 Sqn 'A' Flight, the incumbent flight in theatre. He and some of the lads from 27 Sqn had come to pick us up from the eastern end of the airfield in the battered old green Land Rover Safaris that we had been driving in Prelim Ops.

'Hello, mate,' said Digs, shaking my hand.

I looked forlornly at the piece-of-shit wheels that he had just rumbled up in. 'Mate, these are fucked up,' I said.

'Welcome to KAF' – Kandahar Airforce Base. 'Smile and wave, boys, smile and wave,' he replied.

'If you have any poo, fling it now,' I said.

Before the UK sent troops into Afghanistan, John Reid, the Defence Secretary, had said: 'Not an angry shot will be fired.' Funny that because we had heard that 18 'A' Flight had been taking rounds and things had ramped up. They had found that things were a lot more difficult. The Paras had been getting involved in more contacts than was initially anticipated, since Digs's flight, 27 'A' Flight, had taken over from 18 'A' Flight.

I climbed into one of the Landies and sat in the back with Daffers. Digs was driving and OC 'B' Flight (my Flight Commander) was in the left-hand seat. I was only half listening as Digs explained to OC 'B' his crewing schedule.

'Basically the rotation comprises two days' IRT and two days' HRF' – Helmand Reaction Force. 'The crews are doing four days at Bastion and then swap out with the taskers. The taskers fly out, come up to Bastion and do HRF and IRT. The off-coming IRT and HRF crews then take the afternoon's tasking and end up back at Kandahar,' he said.

As I looked out of the Land Rover window I could see that Kandahar had grown since Prelim Ops. There were more Hesco Bastion walls, more vehicles, more people, and it had expanded a good 25 per cent. Digs was chatting to my boss, OC 'B' Flight, in the front. 'Mate, if you keep prodding a hornets' nest some hornets are going to come and have a pop at you.'

'I fucking hope there's not too much fuck-arsing around today and the handover is good,' I said to Daffers, det Tourette's in full swing.

The plan was to meet up with Digs and get a state of the nation update at the coffee shop a few minutes away from the accommodation block. The Green Bean was what Starbucks would look like if it had no roof, no one ever dusted and all their furniture was made from pallets. That said, they served the full gamut of high-quality coffee products from frappé to cappuccino and everything in between.

It was the venue of choice for the less sensitive crew briefs and debriefs and (in the absence of a bar) for generally putting the world to rights. While for the aircrew it was seen as a relatively rare treat, there were generally so many of the headquarters staff and less 'operationally focused' personnel there that the Green Bean had gained the nickname 'Forward HQ'.

Digs took the floor and began the handover. He gave us an update on how many aircraft were available, where they were located and what the tasking programme was for the next few days.

Kandahar was the main hub with our six aircraft. One of the spokes of the hub was Camp Bastion with the IRT and

HRF cabs permanently based there. We were operating a rotating push Forward with the IRT and the HRF, suitably crewed in this case with three crews Forward. In addition, there were a small number of engineers already Forward to service the aircraft.

The taskers came out of Kandahar and did tasking en route, normally Lashkar Gah or Gereshk, or they would go straight to Bastion. 3 Para were based in the platoon houses in Sangin, Musa Qala and Now Zad, and if we were to go in there it would be a deliberate op because these locations were hot with enemy contacts. Digs explained that they had been taking the taskers up, briefing at Bastion and then going in to do the job.

I was itching to get Forward because that was where the action was. We were with other guys who hadn't been there before because they weren't with us on Prelim Ops. I was pleased when OC 'B' turned round and said: 'Hammond, as you know the crack you can take your crew Forward first. Giles and Dave, you can also go.'

I rubbed my hands together. *Nice one*, I said to myself.

I looked at Daffers and he gave me a thumbs-up. Dan and Sam looked pleased too.

I was relieved that I wasn't going to be sitting around. The priority was to get ready to go Forward now that I was going to be flying the next day. In at the deep end – perfect. Immediately, I had shit to do: zero my weapons and get all of my theatre familiarization check rides accomplished with the QHIs (Qualified Helicopter Instructors).

Digs said: 'Has anyone else got any points, any questions or anything to add?'

There were a couple more questions. Then I piped up: 'What's the relationship between JHF(A)' – Joint Helicopter Force (Afghanistan) – 'Forward and 3 Para like?'

'Yeah, it's very good,' Digs replied.

'What are the lads like up there?'

'Mike McGinty's a good lad,' said Digs.

In my head I was thinking: *Oh yeah, Mike McGinty.* I knew him from my days at Middle Wallop,

Digs finished the brief and the four us had a bit of a huddle to get our shit in one pile.

'Right, lads, our priority is to get ready to go Forward, get our flying kit sorted and pick up our morphine.'

The morphine is the first thing you are going to administer if someone is in pain. It's stored in two coffins, which are what we call the plastic containers. Everyone in theatre, across all ranks and services, carries morphine because it's your first line line of defence if someone's injured, including yourself. Effectively pure heroin, it will give you a good eight hours of pain-free la-la-land, which hopefully is enough time for you to get recovered in the event of an incident. I carry two auto-injectors in my top left-hand pocket. You inject them directly into your thigh.

'And ammunition,' I said.

'We also need to sign for our pistols, zero our weapons and get all of our in-theatre qualifications done and dusted so we are back up to speed with life in the shithole.'

The picture that was emerging was clear. The situation had changed dramatically in Afghanistan. The honeymoon period of Prelims Ops was over and now the campaign was fully underway. All of the action was happening a hundred miles away and at that moment I was sitting in a café drinking a mocchacino frappé latte hot chocolate espresso.

I couldn't help thinking about the lads on the deck who were sitting there sweating their nuts off, managing on sparse rations and minimal water, having not washed for three weeks. I was ready to go Forward into the thick of the action. It was infinitely better than being stuck at Kandahar.

Sam agreed. 'I want to get out there and do something. Kandahar is already doing my head in. There are so many people that never leave the wire. What do they do?'

'Clearly they spend a lot of time pressing their uniforms,' said Dan.

'On my way to the Green Bean, this twat came up to me and bollocked me for wearing the wrong sunglasses. Apparently, reflective sunglasses are not allowed. Can you believe that it's forty-five degrees outside and bright sunlight and you can't wear reflective sunglasses!' ranted Sam. 'They are taking the piss. I cannot wait to get the fuck out of here,' he said.

The crewmen were on a roll. You have heard the expression 'hell hath no fury like a woman scorned'; well you could exchange 'crewman' for 'woman' and it would work just as well. These old ladies were merciless once they got their teeth into something. They were like a dog with a bone. Not that I disagreed with them. KAF was full of PONTIs (persons of no tactical importance).

'Right, ladies, got that off your chest, have you? Then let's go and get some scran!' I said rubbing my hands together. I was absolutely Hank Marvin.

Chapter 5

That evening I went to the gym with Dan. As we walked through the double doors into the vast cavernous space that housed the gym I realized that our Americans friends had once again delivered when it came to offering services and facilities to their troops. The gym was a Bootneck's wet dream, with every aerobic and weight machine known to man, and a few more.

I have died and gone to weight-training heaven, I thought to myself as I warmed up.

'Gucci,' said Dan as he surveyed the area.

'This place is the mutt's nuts, mate,' he said.

'I need to go to hospital,' I added.

'What the hell for?' said Dan.

'Cos my guns are sick. Now let's go get massive,' I replied.

'Prick,' was his reply.

There were stacks of free weights and rows of running machines. They even gave you towels and cold bottles of water. American troops take their weight training very seriously, although for some reason they all seem to focus on nothing but their biceps. We called them lollipop men because they have huge upper bodies and skinny little legs. We signed in using false names. It was a bit of a Brit tradition, to test the Americans' sense of humour. I was Air Vice-Marshall Sir Kissmecrack and Dan was Colonel Hertz Van Rental.

He was not a happy bunny. We had just come out of a welcome-to-theatre brief and Dan had smoke coming out of his ears. I figured it was best if we channelled some of this aggression into the weight room.

'I could not believe it when they opened the brief with "Just take a look around this room – one of you might die"! What the fuck were they thinking of?'

It wasn't a great opener to the det, I had to admit, and when it was said my toes curled instantly. I thought to myself at the time: *No, don't say that. What are you doing?*

'I think they were just trying to set the scene.' Dan didn't need encouraging. But as a senior officer I had a responsibility to try to keep things on track and not exacerbate any difficulties. I replied as diplomatically as possible: 'We know that guys Forward are taking rounds.'

I go to the gym as a release. If I am in a bad mood, I do some phys and I am no longer in a bad mood. If I don't go to the gym then generally I am in a bad mood. It's an important part of my anger management programme, that and chain-smoking Marlboro Reds. During Royal Marine training, one of the Company Commanders came to brief my young officer batch and he gave me some sound advice, which was: 'If in doubt get your fags out.' Any problems that required a solution usually could be sorted over the duration of a cigarette. Smoking is a release and, yes, it does conflict with my phys addiction but to err is human. It works for me.

When I got back from the gym I squared a few things away. I didn't unpack any more kit because I knew that I was moving Forward in a couple of days. I could not wait.

Chapter 6

Before being cleared to operate, each member of the incoming flight had to complete his TQ, or Theatre Qualification, flights with experienced members of the off-going flight. This involved being familiarized with the procedures and the various helicopter landing sites, or HLSs, and making a number of desert landings by both day and night.

Having only just arrived in theatre, we had to spend some time before we first committed aviation, sorting out our flying and survival kit. Before we got airborne over Afghanistan, outside the relative safety of the wire and armed guards surrounding the airfield, we had to be sure of exactly what our actions would be if we ended up on the ground in hostile territory.

We drove up to the CH-47 in the Landie. The cab stood dormant on the pan waiting to be fired into life. Three other British cabs were lined up alongside it on their respective spots on the concrete hard standing, against the backdrop of Five Mile Mountain, a long ridgeline around 1500ft high. Standing either side of the four Brit Chinooks were around thirty other coalition helicopters: BlackHawks, Chinooks and Apaches. I was looking forward to getting airborne again. I loved my job. I felt privileged to be a pilot, and also to be paid to do a job that I loved so much was something that I didn't take for granted. It is everything that I dreamed it would be as a kid. Flying is an addiction and the cockpit is one office that I never tired of climbing into.

The size of the Chinook never ceased to amaze me. The

CH-47 is a big old beast, nearly 100ft from tip to tip and 20ft high, weighing around 13 tonnes without fuel, and with a rotor diameter of 60ft. The Americans began building it in 1963 and it was called 'Chinook', an Inuit name given by the met men to a warm winter wind coming down off the Rockies. Nowadays it has the slightly more affectionate nickname of 'Big Windy', and that doesn't refer to the crew's bean-eating capabilities, or 'Wokka Wokka' because of the thumping sound of the rotors turning.

On the pan, with its powerful blades still and bowing towards the ground, the Chinook snoozed. It may even seem an innocuous, humble machine, but when each of the two Lycoming T55-L712 turbine jet engines roars into life, banging out 4000 shaft horsepower each, linked by drive shafts that split the power to the front and rear heads, and with a thunderous rumble the blades whirr into life, the power of this magnificent piece of engineering is inescapable. Its two sets of three rotor blades, each one 3ft wide and 30ft long, spinning 225 times a minute or nearly four revolutions per second, act as a wing to keep the cab in the sky. Get hit by one of them and it will cut you in half.

This tandem rotor combination results in some tricky design features. In order to help the cab to fly and to taxi on the ground, the heads need to be slightly tilted forward. The forward head is tilted 6 degrees down and the rear head 4 degrees down. This means that the blades of both heads intermesh. It's quite important that they miss each other on the way past.

Many CH-47 pilots, including myself, haven't completely grasped how a Chinook flies. Biannually pilots undergo an assessment by the Standards Evaluation Team, or the 'Sky Police', which includes, among a plethora of other tests, the principles of flight of a tandem rotor helicopter. The answer

34

not being, as I said one year: 'I don't really understand it. I thought it was pixie dust. The engineers sprinkle it on the rotor heads every morning.'

There is a massive hydraulic ramp at the back, which makes it easy to load and unload troops, vehicles, freight and sometimes even boats. As I walked up the ramp, and into the hollow interior of the fortified, rumbling machine, I looked at the grey leather walls, with red seat belts, and was as always awestruck by the magnitude of the beast. The beauty of the Chinook is that it's all cabin. Having no tail rotor taking up useful space was part of the aircraft's appeal.

In the Vietnam War, our American friends managed to lift 120 people, stuffed into the cabin of a Chinook, out of Saigon as it collapsed to the advancing North Vietnamese armed forces. We don't lift that many people these days but I have heard of RAF Chinooks lifting upwards of sixty passengers in an emergency. The cab was cleared for this, and the rules state that the passengers are supposed to stand facing the rear – like that will save them in a crash! At least they won't see it coming.

Nelly Bausor, a seasoned and capable aviator who had been on the Chinooks for donkey's years, was already in the cockpit. I was pretty chilled because it was daylight and there was good visibility under the bright, beating Afghan sun.

The cockpit is divided by a central upright console that contains all the engine instruments. Much like in a car these indicate how hot the engines are and what the oil pressure is doing, as well as how much fuel is left and how quickly it is being consumed – essential for making sure time to destination can be calculated and vital for transits over the sea or over hostile territory. The big difference between the Chinook and a car is that a Chinook has six blades, six fuel tanks, five gearboxes, two rotor heads and two engines, so a few more dials than your average motor.

Flying around during daylight is nice and easy. Landing Chinooks in the desert – now, that is the tricky bit. The huge rotor blades kick up a world of dust. It happens seconds before the wheels touch the ground. The cab is engulfed in a cloud of sand, everything browns out and there is zero visibility. It plays havoc with our senses and is completely disorientating.

Such a complex machine needs a crew of four to operate it effectively, particularly in a war zone. Up front in the cockpit sit the two pilots, or one pilot and a navigator. Actually, nowadays, with female pilots and for reasons of equality, we have to call the cockpit the 'box office' as well. We tend to prefer flying with two pilots in conflicts so if one takes a bullet the other can fly him to hospital. Most navigators get some piloting skills taught to them by our instructors, just in case the worst happens and I need to nip out for a piss. The aircraft can be controlled from either seat in the surprisingly small cockpit. In wartime, we sit in an armoured seat, with armoured side panels, an armoured floor, and add to that our chest plate armour. We are like a flying armadillo.

Sam and Dan were down the back. They are the two people who really run the aircraft, the crewmen, or loadmasters, as their branch is correctly referred to. Helicopter crewmen (or indeed crewwomen, as there are plenty of female crewmen around but for simplicity I will refer to them all as crewmen) hate being referred to as loadmasters as their job entails so much more than that of their fixed-wing counterparts.

Your average helicopter crewman has to be able to oversee the loading of troops and passengers into the cabin and loads to the three hooks under the cab. There are a great many different types of freight requiring different methods of restraint inside and under the aircraft. On top of that they have to be able to navigate the aircraft when the two cockpit crew are busy flying it. They man the two, sometimes three, crew-served weapons (either M60 machine guns or M134 Miniguns) while giving the pilots a verbal talkdown in height and distance to a, usually, very dusty landing site. They do have goggles to help them with that – it keeps the sand out of their eyes. They have to be able to do all this by day or night, in good or poor weather, and in environments from the Arctic to the jungle and all points in between.

They are responsible for looking out of both sides of the aircraft for threats, tracer, RPGs, obstructions, etc. They are also well positioned to ensure that there is no breakdown in communication between any members of the crew and the passengers, and, in addition, they are the winchmen, winch operators and engineers. Finally they can also be the refuellers. However, ask them for a cup of tea and sandwich mid-flight at your peril. Trolley dollies they are not.

A good crewman is worth his weight in gold and can make all the difference to how well and safely a mission is achieved by the cab and crew. We are a team of four in which crewmen play a pivotal part.

To put into context the dexterity of the UK crewman, the Americans in their CH-47s have a crew chief, a door gunner and a flight engineer. The flight engineer is purely there to engineer the aircraft, the door gunner is there to fire the gun and the crew chief is there to manage loads, equipment and embarking personnel, whereas British Chinook crewmen are skilled in all three roles. Essentially the Americans have three people to do the role of one person.

The number one crewman (the numbers relate to the side of the aircraft they predominantly work on and are based on the engine number – number one engine being the port engine, number two the starboard engine) works down the back left of the cabin and is responsible for the maintenance panel. This is a panel showing all the pressures and temperatures of the various gearboxes and filters that they couldn't fit in the cockpit instrument panels. He also operates the hydraulic loading ramp and the M60 machine gun mounted on the ramp. There is not a lot of protection down the back from bullets and RPGs so this tends to be the least favourite spot to work for crewmen. The number two crewman works up the front right of the cabin and is responsible for assisting the cockpit crew with navigation and radios as well as manning the front M60s or M134 Miniguns.

'Mate, I am not looking forward to the dust,' I said.

'Don't worry. It's always the same when you are back in the saddle for the first time,' Nelly said, reassuring me. 'Just focus on landing the aircraft, holding it steady, keep talking to the guys at the back and you'll nail it.'

I checked the inverter, and made sure I had '6-2-3-6' on the Caution Advisory Panel (CAP).

'Clear PU,' said Sam, confirming that the Auxiliary Power Unit, or APU, was ready to fire up.

'Roger that,' I acknowledged then flicked the switch. I counted under my breath as the fuel poured into the APU – one . . . two . . . three . . . four . . . five – then hit 'start'.

I could hear the APU winding up. It began to click over, then the APU 'on' light lit up and the CAP light came on. The sound in our ears changed tone – a reassuring confirmation that all was working as planned.

'APU gens on – PTUs one and two ready panel,' I shouted.

Sam pressed the maintenance panel test button at the back of the cab and six more lights flashed on the CAP.

'Six on,' Nelly reported, then 'Six off' as he released the switch again.

Nelly reached to his right to turn on the kit in the central console: radios, nav kit and the cab's self-defence suite.

'Fire warning test.'

Everyone was momentarily deafened as the sound of the klaxons assaulted their ears.

'Loud and clear,' came the laconic reply.

'Clear the dash.'

I checked the flight controls were moving freely and started the number one engine. The powerful Lycoming turboshaft engine spooled up and, once the noise behind me had stabilized, I engaged the rotors with a shout of 'Clear rotate.' 'Clear all round to rotate,' came the reply from down the back. The big blades started to move. Lazily at first, but faster and faster until the thumping sound created a smooth, loud, regular beat.

'Clear flight on one, please.'

'Clear flight one,' says Sam.

I smoothly advanced the throttle on the number one engine to bring the power up as the blades meshed into a hypnotic blur of light and sound.

After advancing the throttle to flight on the number one engine, I repeated the start process for the number two engine, did the arming checks and it was time to taxi.

'KAF Tower, this is Footloose 25, six on board, holding at Mike ramp. Request taxi to Foxtrot for a westerly low-level departure.'

KAF Tower replied: 'Footloose 25, you are cleared to taxi to Foxtrot, caution, "heavy" landing on 27 and 3 Rotary transitioning to the south. Visibility is five kilometres and the wind is Two Four Zero at six knots.'

'Taxi Foxtrot,' we replied.

Once we had ground-taxied to Foxtrot, the main taxi way that ran parallel to the runway, we were ready to lift to the hover and transition into forward flight.

'Tower, Footloose 25 at Foxtrot, ready for departure as briefed.'

'Footloose 25, Tower, cleared for take-off, caution birds.'

This meant we were all eyes on stalks looking for birds in our departure direction. If you hit a bird at speed in a helicopter it can smash windows, break engines and damage blades. Add that to the other aircraft and enemy threat and we were very focused.

'Take off, Footloose 25.'

I pulled out the master arm pin, called out the pre-take-off checks and then began pulling up on the collective with my left hand. This causes the blades to move up or down together, or 'collectively', and results in the aircraft climbing or descending, by increasing or decreasing the amount of lift generated. I felt the exhilarating feeling that comes from pulling the power and I could feel the aircraft responding as it strained upwards from the ground. The wheels went light and we were airborne.

Once we had lifted to the hover, we did a quick set of after-take-off checks, made sure we had armed up the self-defence systems and then transitioned away at low level into the relative safety of the wide-open Afghan desert.

We had transited out to the south-west of Kandahar towards a river valley. KAF was a very busy airfield with a heady mix of helicopters, heavy transport planes and fast jets. It made the airspace around it very complicated and congested, as well as quite dangerous. Add in the threat from the Taliban and departing was always quite challenging.

The area we were headed for was just on the edge of the desert, where a dried-up riverbed, or wadi, meandered through. There was a lot of dust. It was perfect for practising scary dust landings.

I held the cyclic stick – similar to a gaming 'joystick' – in my right hand. It is called the cyclic as it changes the pitch on each of the spinning blades 'cyclically' as it passes through a particular position relative to the head, and causes the disc of spinning blades to tilt, thereby controlling the direction of movement of the cab. Push the stick right and the helicopter moves right, etc.

I had my feet on the yaw pedals. On the Chinook these are really expensive footrests as we hardly ever use them unless in the hover. The pedals are there to spin the aircraft around a central axis – in the Chinook this can mean spinning the cab around the middle, the nose or the tail depending on what you want it to do. Add in a bit of throttle, some flying controls, spinning blades and, hey presto, a flying helicopter.

As the Chinook has such massive blades and is a heavy beast it needs hydraulic assistance to power the flying controls, without which it would be impossible for the pilot to move the blades and control the cab. There are two hydraulic systems in case one leaks or breaks, and a final back-up system to the hydraulics should they both fail, which makes it designed to soak up bullets and yet still get the crew and passengers home safe.

As the aircraft descended we entered what is called 'the gate'. I was down to 100ft and I brought the speed back to 30 knots.

Nelly Bauser said: 'You are in the gate' and he set the radar altimeter alarm to 40ft.

I then set a decelerative altitude and started my descent by lowering the lever with my left hand. Dan began to patter the dust cloud at around 80ft.

'Dust cloud forming,' he said.

My descent continued.

'Dust cloud at the ramp . . . centre . . . door.'

As I was coming down I could see the dust cloud coming round towards the nose of the aircraft. If I was at 10ft by the time it got to the nose I knew I would be all right. I could move forward and down, and then land on the back wheels.

The Chinook has sand filters, the EAPS (engine air particle separators) attached to the front of the engine intakes. They are like large dustbins and evacuate the sand through enormous suction pumps and big sand fans that separate the sand from the air. Sand is the enemy of everything. There are very few things that sand agrees with. For high-speed rotating components like rotor blades and gearboxes it is a grinding agent, quickly wearing down the brittle metal parts. It erodes everything and gets everywhere. Each aircraft spends nine months in theatre before returning to the UK. Even though they have been washed, cleaned and hoovered inside and out we still get 400 kilos of sand out of them when they go into deep maintenance and the floors are lifted. It's like when you go on a beach holiday – no matter how vigilantly you clean yourself after a day at the seaside, somehow you still end up with crunchy sandy sheets in your bed at night. Sand just gets everywhere.

I had problems doing dust landings in the Chinook because I always thought like a tail rotor guy from my days flying the Lynx and the Cobra. It was a real change of mindset for me. If I came in like that on a tail rotor I would rip it off. Coming in on the rear wheels was therefore like getting a new cassette in my head.

Nelly Bauser knew this and kept me focused on what I needed to do.

'Don't fight it, mate. Keep the nose-up attitude. It's tandem not tail rotor. You've done it before,' he said.

My muscle memory wouldn't let me drop the tail because I had flown helicopters with tail rotors for over ten years.

As it was during the day I had the peripheral vision I needed and could see all around me. I held it on the rear wheels, kept the collective in a bit to cushion us on to the ground and then lowered it down.

'That's it. Great job, mate,' he said.

'All wheels on,' said Sam.

'One down, many more to go,' I said.

I had cracked the first one and I felt a great sense of achievement.

'Right, same again,' said Nelly.

We spent the next hour flying mini-circuits to landing and repeating the process until I felt like I was all over day dust landings. As we landed back at KAF, I turned and looked at Nelly. 'Just the nasty night ones now and we are laughing,' I said. That would be nervous laughter.

That night I went out with Pete Ward, another seasoned QHI, who had been on the Chinook twenty-odd years man and boy. Over the last year Pete had been instrumental in working me up to combat-ready status. He had schooled me in the complexities of tandem over tail rotor flying. This trip was notably scarier than the day one.

Night dust landings using night vision googles – NVGs – were the sport of kings. It was hard work. Goggles work by image intensification. There is a scale from zero to a hundred which measures ambient light, with zero being dark and a hundred being very bright.

Ambient light, whether star- or moonlight or the lights of towns and houses, is processed as electrons through a photocathode tube – one of two attached to the front of my flying helmet. As the electrons pass through the tube, similar electrons are released from atoms in the tube, multiplying the original number of electrons by a factor of thousands through the use of a microchannel plate in the tube, intensifying the image and projecting it as a green view of the outside world in front of my eyes.

But, really, wearing NVGs was like looking through two green toilet tubes weighing half a ton that had been attached to my head and rammed up against my mark one eyeball. Until you get used to them, they can make flying at night quite interesting.

Dan, Sam, Pete and I fired up the cab and headed back out to the riverbed. By night, lifting from KAF was even more dangerous as a lot of inbound aircraft did not show any lights for fear of being shot down by the enemy. This made them really hard to see and avoid. We got to the same wadi and set up in the gate for our landing. Night dust landings are exactly the same as day ones – at least, that is the principle. In reality they take a bit of a leap of faith and a lot of training. There's no peripheral vision on NVGs but if I set up in the 'gate' at the correct height and speed, then set the right attitude and rate of descent, theoretically the aircraft should progress serenely down towards the ground, defeating the laws of gravity, battering the air into submission and land itself.

The problem is, when people are shooting at you, serene approaches are nowhere to be seen – that is where the training and leap of faith come in.

In the cockpit we have the instrument panel and flying controls – to many people a bewildering array of dials and switches and flashing lights. The controls are mirrored so either pilot can fly, but most of the nav kit and self-defence switches are biased towards the left-hand guy. This set-up works well and allows the pilots to concentrate on their own parts when operating the aircraft. The dials on the panel in front of the cockpit crew mainly relate to flying the cab and keeping it in the air.

There is an attitude indicator, or AI, which is an artificial representation of the horizon in the real world. It allows us to fly the cab in the dark, clouds and bad weather solely by looking at the instruments. This is an environment we are comfortable with but intensely dislike. As helicopter pilots we should be down in the weeds, skimming the tree tops and swooping though valleys, not sitting in the middle of a dark sky, staring at dials two feet in front of our eyes.

Underneath the AI is the horizontal situation indicator (HSI) – more commonly known as a compass. This compass has many different dials superimposed on it to indicate the direction of different types of navigation beacons and information about the Instrument Landing System (ILS). It enables us to fly an approach to an airfield in the dark, quite similar to regular passenger jets.

On the right-hand side of the AI is the radar altimeter, which gives the radar height of the cab above the ground as it passes beneath the aircraft and is vital for flying at extremely low level and for landing in the desert in very dusty conditions. On the left of the AI is the airspeed indicator (ASI), so we can see how fast the aircraft is flying through the air, in nautical miles per hour. Below the RADALT is the barometric altimeter, or BARALT, communicating the height of the aircraft above mean sea level, or above the ground when flying at higher altitudes. Below the BARALT is the vertical speed indicator (VSI), providing information on the rate of climb or descent of the cab, in feet per minute. Finally, to the left of the HSI is the standby AI, a back-up instrument if the main AI fails when we are in cloud. These instruments rule pilots' lives, especially when we are engulfed in cloud with no visibility. My eyes scanned them continuously to check all was well because I couldn't see the real world out of the windows.

In order to hold the cab steady I set the power on the collective and manage the symbiotic relationship with right hand and left hand. Being a pilot is a matter of co-ordination. I tended to look outside and see how fast we were moving across the ground so as to judge whether I had got it right or not. I was constantly checking to see whether the sight picture approach was right.

At the point where what little of the ground I could see on NVGs disappeared into the all-encompassing dust cloud, when I was still a few feet up in the air, I had to hold everything where it was and hope it would all work out ok. Most of the time it did.

I was ready to bring the cab down. The dust cloud had swallowed up the base of the aircraft. At the start of the approach I could see very little – it was too dark and I had peripheral vision spanning six inches either side of my head. The only way to land the aircraft was by looking in and relying solely on the instruments. By day your peripheral vision provided loads of clues about speed and any drift left or right, a luxury lacking in the dark – hence the leap of faith and the necessity of being set up at the right speed, height, rate of descent and aircraft attitude. In the final stages of the landing it was pitch black. A deep velvet blackness without an end or a beginning. I put all my faith in the instruments. If it went wrong I would rip the wheels off. The pressure was on to do it properly. I had to get this nailed. I was about to go Forward and there would be much more at stake.

'Wheels down,' said Sam.

'Great job, Mark,' said Pete.

'Let's go again.'

We spent the hour going over and over the dust landings. It didn't get easier and even after doing it again and again and again, all I came away feeling was I needed to do it more. But that was my lot and we were now ready to go Forward.

Chapter 7

As we walked on to the back of the C130 Hercules, destined for Camp Bastion, I was savouring my passenger status, no planning, no effort. It was a relatively painless exercise as all we had to do was pitch up at the back of the aircraft, dump our kit and sit down onboard to enjoy the quick trip Forward.

We lifted at 0830 so I had packed the night before. I had loaded my bags with essential items: head torch, flip-flops, dobie dust (shower gel), book, towel plus my standard military stuff. I took my kit up to the aircraft and made sure it was all aboard. I had to make doubly sure that I had body armour and a helmet. It would have been more than inconvenient to arrive into Bastion without it. Being in a war zone without your body armour and helmet is like driving a car without a seat belt – you can do it but it definitely increases the risk of death!

All three crews going Forward – mine, Giles's and Dave's – boarded the C130. The inside was the same as a CH-47 but a lot bigger. It had the same red sling seats, drab colour scheme and a bewildering array of pipes and cables along the roof and walls. A Hercules operating on a passenger move basis will have four rows of seats, one on each of the outside walls and a double row back to back down the middle, facing outwards. If there are not many people onboard it can be quite comfy, but pack it full of passengers and kit and it can be a total nightmare. The noise is fairly bearable and they tend to smell of the usual heady war zone mix of hydraulic fluid, aviation fuel, fart and BO. It's the opposite of the Lynx effect. However, they were great old birds, robust and solid and pretty dependable. A real workhorse and a very flexible aircraft – almost as good as a Chinook.

Giles knew the pilots. They offered him and me a seat in the cockpit for the journey up. After take-off, we clambered up the steps and I sat on the jump seat between the pilots. I surveyed the landscape. It felt more remote, infertile and hostile from the Hercules's greater height. In the distance I spotted the Red Desert and I thought to myself: *Yeah, I am definitely back.*

The Red Desert is an arid area of south-western Afghanistan. It is a vast expanse of desolate space with red sandy ridges and small, isolated hills. The red sand dunes and ridges reached heights of between 50 and 100ft, and are sandwiched between windblown sand-covered plains, bereft of vegetation, interspersed with beds of barren gravel and clay.

I had done this trip a hundred times during Prelim Ops. A slight sense of trepidation niggled me. I didn't show it, and I didn't share it. We were going Forward to a place where the aggression had increased twofold. My Chinook mates had been shot at, which meant we would be shot at. Before it was a milk run and now it felt different.

What most military aviators suffer before they go into a hot conflict zone is not fear but performance anxiety or stagefright. We have trained our entire professional lives for this day and to walk out on to the field among our peers and the guys out there fighting for their lives and to fuck up would be the worst thing that could happen.

The fear comes later. When you get back. Occasionally there are moments in everyday life at home and there's a flash in your head and you think: *Was that real?* It was almost like it had happened to someone else or I am so removed that I think it's a war movie. It's fast and jumbled and at the time I was so focused on delivering the task that I didn't have the capacity to observe what was going on and absorb the experience.

The C130's tactical approach into Bastion was a new experience for me. At height they throttled back the engines, then dropped the nose hard down in order to get through the threat band quickly before levelling at low level just short of the runway. The pilots stood the aircraft on one wing, keeping the enemy guessing as to their next manoeuvre. My stomach lurched as the G-forces came on.

I realized there was a huge amount skill required by the pilots to haul nearly 40 tonnes of a four-prop workhorse around the sky and drop it into a landing like it was a Cessna. The guys on the sticks made it look like a piece of piss.

At low level I could see that where there once was a wide expanse of desert now stood a city of tents, a concrete runway and hard landing pads for helicopters. None of that was here when we left last time. Before it was just dirt and a few tents. Now it had expanded into this massive, fixed, operational, working war machine. The landing was hard – this was no BA flight.

Camp Bastion was the main British military base in Afghanistan and could be found to the north-west of Lashkar Gah, which was the capital of Helmand Province. The camp was around four miles long by two miles wide, with an airstrip, a field hospital and stacks of accommodation. Stationed at the camp were around 3000 British, Danish, Estonian, Czech and American troops. Helmand was the main source of Afghanistan's opium output and was in the grip of a Taliban insurgency.

During Prelim Ops I had been insatiable in my reading about the history of Aghanistan and it didn't paint a pretty picture, but I wanted to understand the intricacies of the story so that I could see how we had come to this point.

Geographically Afghanistan is situated at the crossroads of Central Asia, connecting the pre-Soviet Russian empire with the Indian empire. Since as far back as Alexander the Great, Afghanistan has been invaded by everybody and their dog. The British have been there several times, most notably the Russians have been there, and in fact Genghis Khan and the Mongols, who subjugated Afghanistan in the 1200s, are the only people ever to have done so in its warring history.

The country's borders were laid down by the Russians and the British in order to keep their two areas of interest apart. They took no account of any ethnic or geopolitical boundaries.

Afghanistan was designed, and allowed to exist, in what was known as 'The Great Game', the period of strategic rivalry and conflict from the Russo-Persian Treaty of 1813 to the Anglo-Russian Convention of 1907.

The country was used as a buffer state between the Russian and British empires. There was an unspoken agreement between the two Imperial giants that as long as they didn't step into the territory of Afghanistan, the Russians would never get to the Indian border, and the British would never expand into Russia.

Afghanistan was initially created to be unstable and it is this founding principle that pervades every part of Afghanistan's existence today. As a country it remains wholly unstable.

It continues to be in the interests of several countries for Afghanistan to be perpetually destabilized – of Pakistan, Iran and of Russia too. Having this hornets' nest buzzing in the middle of them keeps a lid on some of the other underlying problems, such as radical Islam and tapping the natural gas and oil resources of the southern Asian republics, which exist in the modern political climate. I was fascinated to learn that the seeds of today's problems appear to have been sown in

1973, when the king of Afghanistan, Mohammad Zahir Shah, was overthrown by his brother-in-law and cousin, who proclaimed the end of the monarchy and formed the Republic of Afghanistan. It began a violent period of coups, killings and instability that culminated in the Soviet invasion of 1978.

The presence of 115,000 Russian troops was instrumental in the rapid growth of the 'mujahideen' ('soldiers of God'), a loose alliance of every faction that opposed the Soviet invasion.

During the quieter times of Prelim Ops the guys and I would often sit around and discuss Afghanistan politics. We had all started reading into the history of Afghanistan so that we could better understand the enemy that we were fighting. Every professional solider wants to know more about the enemy and understand his motivations. As I delved deeper into the story I realized that the situation that we were facing was far more complex and difficult then I had ever imagined.

To counter the Russian invasion the mujahideen were heavily funded by the West, in the battle to prevent the spread of communism at the height of the Cold War. At the same time as receiving this Western backing they were also being significantly influenced by extreme Islamic groups, including one organized and financed by Osama Bin Laden. It attracted volunteer Muslims from other countries to assist the various mujahideen groups in Afghanistan, and gained considerable experience in guerrilla warfare.

In 1989, the mujahideen were victorious. The Soviet Union withdrew all of its troops and its puppet regime was toppled. Yet the mujahideen did not create a united government in its place, and infighting for power in Kabul ensued between many of the larger mujahideen groups.

The resistance movement of Afghanistan was born out of anarchy and a civil war raged. The effect was devastating and lasted several years until a village mullah, with the backing of Pakistan, organized a new, armed movement called the Taliban.

By 2001 the Taliban had largely defeated the militias and controlled most of the country. Their main spiritual power base was in Kandahar and it was from here that they controlled the country. They never governed centrally from Kabul and attached little importance to the capital.

They ruled through protectionism. Their financial system wasn't administered via a central government but seemed to be run more according to the law of the jungle, in which the strongest ruled. The locals would pay tax to the local strongman, and whoever was richest could afford the biggest army and therefore got richer and retained power. They imposed a brutal, hardcore, radical Islam on the beleaguered Afghan population. And, eventually, the rest of the world found out all about it. Spectacularly and violently.

Chapter 8

On 11 September 2001, I was flying in an RAF air transport to Cyprus to join HMS *Ocean* for Exercise SAIF SAREA. I was meant to be out for only eight weeks, but as a result of 9/11, and being the Lynx Flight Commander on 847 NAS, I was nominated to be the detachment commander for the aircraft that were going to stay and support operations in Afghanistan. This was onboard HMS *Fearless* in the Gulf of Oman. I didn't return home until six months later.

The attacks on the World Trade Center came as a shock to everybody on this detachment. We stayed out in the Gulf of Oman and witnessed from a distance the US-backed victory of the Northern Alliance, an alliance of the remaining non-Taliban-aligned mujahideen factions over the Taliban.

This defeat had come about owing to the Taliban's unwillingness to give up Osama Bin Laden after the 9/11 attacks on the World Trade Center. He had been tried in absentia by the US military courts following the East African embassy bombings, which put Bin Laden on the list of the world's top ten most wanted criminals. The charges against him were that he created and financed a terrorist organization, al-Qaeda (meaning 'the base'), a term seemingly coined by the Western governments to put a face to organized Islamic terrorism.

Osama Bin Laden had set up terrorist training camps in Afghanistan, with the tacit agreement of the Taliban. It was in these camps that the likes of the 9/11 bombers were trained. Even as early as 1998 these camps had been targeted by the US military. It came as no surprise to us that without the release of Osama Bin Laden to the West, the full might of the US military was going to be unleashed on the Taliban.

The West declared that Afghanistan was no longer a safe haven for terrorists. Then the Northern Alliance were slowly merged into the new national army and police forces, or demobilized.

In the middle of the US-led usurping of the Taliban, in December 2001, the United Nations brokered the Bonn Agreement, creating a blueprint for the development of a new government in Afghanistan and tying Great Britain into the mission of stabilizing the country.

In 2001 we succeeded in depriving international terrorists of a refuge in Afghanistan but then set ourselves the monumental task of creating a viable state in that country, which we have not yet achieved. But we are not going to give up trying. Sometimes we take two steps back to make one giant leap forward.

Chapter 9

Bloody hell, I thought to myself, *this place has grown*. Whereas before Camp Bastion was just a couple of tents in the desert, now it had expanded into a mini-town. It was the main hub of British forces in Helmand and there was a real buzz about the place.

It was good to see the Apaches were here. The Apaches weren't deployed during Prelim Ops. We had actually put little tags at key points at Lashkar Gah and Gereshk on our last visit which stated 'certified safe for Apaches, love 18 "Bottle" Squadron'. But now everyone was at the party and the party was a fully spun-up war-fighting machine.

The Apache force had first deployed at the end of April 2006 and was permanently based at Camp Bastion. It had provided the majority of the infrastructure, the people, the kit and the command chain of all of the aviation platforms at Camp Bastion, which was managed under the JHF(A). At this time JHF(A) was commanded by an Army CO, Col. Richard Felton, who was located back at Kandahar, in the Rear headquarters.

Bastion was in the middle of the desert. It's usually the best place to put a Forward operational base. The entire camp was surrounded by a large berm, or an earthen wall, with guards patrolling the perimeter. The berm provided an obstacle to vehicles, particularly armoured fighting vehicles, but could also easily be crossed by infantry. The land around Bastion was flat and the camp was strategically very well positioned – it was in the middle of nowhere. If the Taliban tried to whack it they would get malleted. It was like Camelot. It was unapproachable even by road because the dust cloud from vehicles would be visible for miles.

The Province of Helmand was made up of districts and these districts all had centres. The District Centres were the local home of the national government in each area. What were to become the platoon houses were often the governor's house or the district chief's house and policed as a government house. It was a frontier post, a little outpost of the Afghan National Government in that area, so politically it was very important to understand that if we weren't there and the Afghans weren't there then the Taliban ruled.

Even if the only bit of land that we owned in a town was 50m square, we were flying the flag, which meant the Taliban didn't own that town. There was definitely a strong political imperative from Kabul to get an Afghan National Government presence alongside British soldiers, literally and figuratively flying the flag for the ANG. However politically significant this was, the platoon houses had received some heavy shit by the time we arrived in theatre and the situation was not improving.

There was no quick-fix solution for the International Security Assistance Force (ISAF) troops on the ground. We were trying to stabilize a region of tribal peasant farmers, questionable government officials and war-mongering Islamic extremist Taliban insurgents. It was no easy task.

It was a counter-insurgency campaign rather than a war. There was enormous consent among the population, political leadership and the people on the ground for us to be out there and to do what we had set out to achieve. There were daily accounts of major skirmishes or tactical actions, which to the man on the street looked like a war. But, in reality, when you got on the ground in Afghanistan there was a lot going on that was really peaceful and positive. However, when the enemy was encountered on their terms, at their time and place of choosing, it was very kinetic and there were bullets flying everywhere, which was dangerous for the civil population.

Prelim Ops had been like a party that was just getting started. We had arrived early at 6 p.m. and no one else was there, apart from the guy setting up the disco. However, arriving in Bastion this time the party was in full swing. Back in the UK I had been asking myself whether it was going to be as good as everyone said it was. But it was much bigger and better then we had ever anticipated. It was now 10.30 p.m. and it was pumping.

As I walked away from the C130 still turning and burning behind me, I dragged myself out of that surreal moment and looked around. *We're not in Kansas any more, Toto*, I thought.

Chapter 10

While we were waiting for our pick-up, I was taking a moment to enjoy a Marlboro Red. I looked up and could see Sam giving a mover a chest poking.

The RAF mover, a person who puts bags on and off aircraft, had cut the sleeves off his shirt, and was wearing a bright red bandana, wrap-around sunglasses and a hell's angels moustache. He had run up the C130's ramp and screamed at us in the style of a Regimental Sergeant Major:

'Right, HOT LS – get moving and get your kit off!'

We were all loaded up with tons of kit. We had our own personal kit, our flying kit and some stores, as well as our weapons. We had so much stuff that it took two runs to unload it. Sam had his arms full to the gunwales.

This REMF (a rear-echelon motherfucker, someone who works well away from the action but is full of unjustified self-importance) in the bandana shouted: 'Fucking get your kit off!'

Sam replied: 'That's what I am doing, you prick, but my arms are full.'

The mover yelled at him: 'I said, fucking get your kit off!'

'I will once I have put this down. I am not a fucking octopus.'

After the aircraft had taxied off, the REMF came over to Sam and started shouting at him: 'When I tell you to get your kit off, you get your fucking kit off. This is my aircraft when it's on the ground!'

'Oh really,' says Sam, 'did you sign for it, did you? Did

you sign the 700, did you? Are you responsible for it if it crashes?'

The REMF looked at Sam and went a little pale. Clearly he hadn't thought he was aircrew.

'No, I don't think so, you are just some fucking idiot who likes to have a go, now piss off!' said Sam. 'Hot LS . . . this place is in the middle of nowhere, twat!'

This was not unusual behaviour from an REMF. It's the product of guys who mainly don't go anywhere. When they do get posted and get a sniff of some kind of action, they take full credit for it, have a total personality bypass and turn into cocks. The military is packed with people who don't actually do much but who all want a piece of the action, and the medals.

The REMF stalked off. I wandered over to Sam, who was standing with Dan. 'What the fuck was all that about?' I asked. Sam explained. It really pissed me off. 'Do you want me to go and tear a strip off him, mate?' I asked eagerly.

'What are you going to do?' asked Dan smiling.

'He's going to walk over and be large and imposing, aren't you?' said Sam.

True, that had been the extent of my plan.

'What a cock,' I said.

The guys from 27 Squadron pitched up to collect us and to take us to the IRT tent from the airfield, in the same battered old Land Rovers that we had left there six months ago. Adams Watts was driving.

'Nice motor,' I said. 'Is it new?'

'Yeah, right,' he said laughing.

I threw all my stuff in the back, as did Daffers, Sam and Dan, and climbed into the front seat next to Adam. The others crammed into the back. He did a handbrake turn and accelerated out of the airfield. It was like *Wacky Races* with clouds of dust flying as we sped into the beating heart of the war machine.

As we drove along I realized that Camp Bastion was outside the realm of anything that I had ever experienced in Afghanistan before. It was different, it smelt different and there was a frenetic hustle and bustle about the place. We had entered a Forward Operating Base in a fully operational war zone.

Even though it was still relatively safe, there was a definite sense of pressure. The only other time that I had felt this level of tension was when I was in Umm Qasr, across the Kuwait border in southern Iraq, in Gulf War II. The 15th MEU, the US Marines, had just taken the city. We had landed and even looking around I could just tell that it had all been kicking off.

As we drove through the camp, there was a tension which was almost tangible. There was no downtime here. People were running about doing their business. There were no leisurely briefings over a café latte at the Green Bean café. I could see the strain in the faces of people as we drove past them. They looked stressed. Their brows were tight. There were few smiles.

I was aware that we were the new guys on the ground and that we had a lot to learn. The younger guys that were with me appeared more startled, and looked like deer caught in the headlights. We didn't say much to each other. We just looked around and drank in the atmosphere, and quite a lot of dust.

As we drove down the main drag, we parked up outside the IRT tent to drop off our kit. We offloaded our bags into the tent, which was one of eight inside a Hesco Bastion-surrounded pod. The tents were air-conditioned with eight camp cots, hanging shelves, black plastic non-slip flooring, a free-standing fridge freezer and a TV in the corner. This was the home of the IRT and HRF crews, with the other crews living in similar tents within the same pod. Across the road was the JHF(A) Forward tent and then next door to that was 3 Para's CP.

Chapter 11

We were supporting the platoon houses in Sangin, Musa Qala, Kajaki, Now Zad and Garmsir as well as the MOGs – the Mobile Outreach Groups. These were new landing sites for us because all we had done before was Bastion, Lashkar Gah and Gereshk. If anyone was injured or there was a call for reinforcements, those were the sites we had to go into. The first thing we had to do after ditching our bags was talk to the guys on 27 Squadron, have a look at the imagery and find out what the set-up at Bastion was.

We went straight to the JHF(A) tent, where I found Major Mike McGinty, OC Forward. He had chiselled good looks and a foppish sweep of hair, and was an all-round 'bloody nice bloke'. McGinty's right-hand man, or Ops Officer, was my mate Lt Cdr Dave Westly (RN). We had served together on 3 BAS and he was fully aware of my attack helicopter experience with the US Marine Corps, flying two-seat AH1-W Cobras. It paid dividends with the Apache guys because they knew I understood where they were coming from.

I also knew some of the Senior NCOs, including a Warrant Officer I did my pilot's course with in 1993 and hadn't seen since. I walked straight up to him and shook his hand. 'Hello, mate,' I said.

'How's it going, sir?' he said. 'I haven't seen you since we were on the course together.'

'Not bad. How long have you been on Apaches?' I asked.

'About three years now.'

We then had a quick chat about the other guys on the course. I realized that he was the only one who was currently on Apaches in theatre from that time. But the CP was full of mates and that made me feel better. I realized that there were a lot of people I knew flying Apaches. It was comforting to know that I already had an established relationship with lads I was going to be working with over the next few weeks. I was a known quantity to them. They knew what I could do and also understood my sense of humour.

Bootneck culture is based on banter and taking the piss. To non-Bootnecks not familiar with merciless, never-ending, having the piss ripped out of you style of banter, it can come across as a tad full-on when you get seen off.

When you get seen off it'll be a stitch-up that pisses you off but at the same time manages to be fucking funny. It's hardcore and childish but there you go. It is part of the Bootneck mentality. We learned it from our very first day in the Corps. On the very first night of Bootneck training, the course above you, who are not qualified marines, come down and haul you out of your beds to thrash you by taking you out on some crazy arse night run. We didn't know that they were recruits, because if we had we would have told them to fuck off. It was my first experience of a classic Bootneck stitch-up.

Another classic Bootneck gag is for someone to surreptitiously put heavy weights into your Bergen when you are not looking. You get into the field having done a mega insertion yomp of 40k, go to get your sleeping bag out and find that someone has put a massive iron bar weighing an extra 10 kilos in your bag and you have carried it all the way. Fucking hilarious!

Banter aside, the guys in the CP were good lads. There would be no fannying around or beating around the bush and we wouldn't have to waste time getting the measure of each other. I knew it was going to be straight talking, no bullshit, and that was how I liked to operate.

Me, Dave and Adam Watts from 27 Squadron briefed in the CP with the rest of the 27 Squadron lads. It was imperative that we began an extensive handover so that we could learn where all the landing sites were. We got out all the maps of the area and started examining them to establish the new, to us, positions.

We were sorting out how we were going to orchestrate the next few days and who was going to cover which duty. We agreed we needed to see first hand where we would be going. I turned to Giles and said: 'We need to get a good handle on these landing sites. I think it would be best if we just put our aircraft captains with the 27 Squadron crews over the next forty-eight hours.'

It was agreed that Dave would cover IRT first, then that Giles would take HRF and I would be the spare.

It was clear that that we would soon need to organize a rotation system of how many days we were going to cover on each line. I wandered over to have a chat with Mike McGinty.

'All right, mate,' I said. 'Long time no see.'

'Yeah, it's been a while,' he said.

'What was it, must have been Middle Wallop in '99?' I said thoughtfully and then asked: 'How is it up here?'

'Busy. It's pretty full-on,' he answered.

'How's the set-up?'

'It's good, we're working well with 3 Para.'

'Yeah, I had heard that from 'A' Flight. So what's occurring? How's the CO?'

He nodded his head. 'Tootal's doing ok. He's a good man.'

McGinty worked alongside our OC. They were responsible

for the tactical operation of Apaches and Chinooks at Bastion. However, when it came down to it, the ultimate decisions were taken by the captains of the respective airframes. My boss was responsible for 1310 Flight, the name given to the UK Chinook detachment in theatre, which was made up of aircrew, engineers and support staff.

Chinook crews deployed to theatre for a shorter period than the Apache aircrews. In order to create a truly joint helicopter force a good understanding of the working practices and differences between the Army Air Corps and the RAF was required. Both had a distinctive style, which as a Royal Marine working with the RAF I could generally understand. Mike was also responsible for his Apache squadron and its tactical employment.

The whole system worked very well. Inevitably there had been a bedding-in period but that had happened with the squadrons before us. It's like a lot of dogs wandering around sniffing each other's arses, but now things had settled the working practices were tried and tested and everyone on the whole was happy with how it was running.

The overall flying programme for the Chinooks (i.e. who was going to fly from either KAF or Bastion) was run from KAF by OC 'B'. The Chinook tasking system was based on two aircraft Forward, one being IRT and the other HRF. We provided three crews in order to cover IRT and HRF, twenty-four hours a day. Back at KAF there would be two general ISAF tasking aircraft. This meant we needed a system to provide IRT and HRF at the required notice to move, while ensuring the crews remained inside their crew duty period of fourteen hours on and ten hours off.

Although Mike McGinty was responsible for the AH – attack helicopter – flying aspects of Bastion, the Command and Control, or C2, was more complicated. JHF(A) was not solely assigned to the British Forces in Afghanistan. It was a NATO, or ISAF, asset. The Headquarters Regional Command South, or RC(S), was under the command of a NATO two-star General, based in Kandahar. This HQ held the command responsibility for five provinces in Afghanistan: Helmand, Kandahar, Oruzgan, Nimruz and Zabul. The majority of British forces were based in Helmand Province and while the Apaches at Bastion and the IRT and HRF Chinooks predominantly supported those troops in Helmand, the other UK Chinooks could be tasked at any stage further afield.

RC (North) was under the command of the Germans and RC (West) under the Italians. RC (East) was the Americans' area, but was not part of ISAF. It was managed under Article 51: self-protection to root out terrorism, whereas ISAF was there under the invitation of the Afghan government.

RC(S) was essentially Brits, Dutch, Canadians and some Americans thrown in for good luck. The Romanians were in Zabol, the Canadians were in Kandahar and the Brits were in Helmand. The Command of RC(S) rotated between the countries.

Within Helmand there was a one-star headquarters, the British forces headquarters. At Camp Bastion was the 3 Para Battle Group under the command of Lieutenant Colonel Tootal and the Bastion Support Group, who were under the control of Helmand Task Force, based in Kandahar.

In theory, whatever McGinty needed to do as OC Forward had to be cleared by and released through NATO headquarters. It was a political statement from the British, asserting that helicopters had been sent to Afghanistan to support NATO, not just our own national interests. Our helicopters along with American and Dutch helicopters were supporting Canadians and many other nations.

The principle was that JHF(A) headquarters in Kandahar

called the shots. Forward in Bastion had to go to them for all clearances and then JHF(A) HQ went back to the NATO headquarters in RC(S) for key decisions, a pretty complicated and convoluted chain of command.

In practice, the helicopters working in Helmand mainly supported 3 Para. The JHF(A) tent was next door to the 3 Para Bastion-based HQ. Whenever there was a big job on for 3 Para, requiring aviation, things would step up a notch. JHF(A) elements would move into the 3 Para CP. There were Forward desks already allocated in the 3 Para HQ because this working relationship was so close.

In the meantime, the specialist flying requirements would continue to be juggled in the JHF(A) tent, so as not to interfere with the 3 Para battles. The aircraft captains, a signaller and a watchkeeper would move into the 3 Para CP, so that we received the relevant information first hand.

Sometimes McGinty and the lads would have to play the game and agree with Tootal to go through the process of determining the 3 Para tasking requirements and then telling the Rear what they thought they should tell them to do. It was quite often a case of the tail wagging the dog.

There was always potential for a disconnect between RC(S) HQ and the Helmand Task Force. Because of the level of activity with 3 Para in Helmand Province, the majority of JHF(A) assets naturally gravitated towards tasking in Helmand.

What mostly happened was that we all just got on and did what needed to be done. The Rear headquarters were very good at recognizing that it was about situational awareness and that JHF(A) Forward had that.

There were always two Apaches and two Chinooks on immediate readiness. They could be called for anybody, not just Brits. The CP was constantly buzzing. While we were in there the MIRC computer system burst into life – MIRC is an American-based computer chat system that allows us to send messages to other units very quickly and prevents some of the ambiguity that can go with radio transmissions.

'*TIC*' – troops in contact – '*standby.*'

'*TIC Musa Qala.*'

'*TIC poss casualties.*'

The CP started to spool themselves up. Questions flew around the room.

'Does it look likely that we are needed?' McGinty said.

The Forward duty watchkeeper sent another watchkeeper over to 3 Para HQ to have a sniff around.

'*Casualty T1,*' clicked the chatroom system.

McGinty shouted: 'Get the Apache crews up here, please.'

As we were looking at the maps, the Chinook IRT crew had already arrived in the CP.

'Yes, sir,' the watchkeeper replied.

'Call the Line,' McGinty shouted.

The Line was the engineering line and it was where the engineers who saw off the aircraft were based. A watchkeeper spoke down the phone: 'Yes, it looks like we are off.'

Suddenly it seemed like everyone had converged on the tent.

In the early 1990s, the Immediate Response Team came into being in Bosnia, with the Commando Helicopter Force and its Sea King helicopters. It developed this role under the UN Protection Force (UNPROFOR) so it could carry engineers, medics and fire teams and respond to any situation, anywhere in the country.

The origins of the IRT were not only with the Commando Sea Kings in Bosnia, but also with the RAF Pumas in Kosovo. However, it eventually migrated to the Chinook simply because of the aircraft's capacity to carry more people and equipment. By being able to carry more kit it had far greater capability in the battlefield.

The IRT was created to define clear parameters for the helicopter crews to be able to respond appropriately to medical emergencies on the battlefield. It became part of the peacekeeping operation to provide immediate-response teams in situ to save life and prevent further injury.

The root reason for the IRT's creation was the large concentration of mines and unexploded ordnance and the primitive infrastructure in Bosnia-Herzegovina, which made it difficult for the armed forces to reach casualties within the time required if they were to save them, the 'golden hour' as it was called.

It became necessary to winch medics down to recover casualties in highly demanding and often very dangerous circumstances. It was a very different approach to a CASEVAC – casualty evacuation – where you turned up at a grid reference, stuck a casualty onboard then flew away. The unpredictable nature of the incidents meant that the IRT was an intensely pressured role.

The pressure is just as intense in Afghanistan, where we know the nature and status of the casualty but we don't always know what the conditions are where that casualty is sitting, be it a minefield, a surrounded and ambushed compound, or even just in hostile terrain. This puts incredible pressure on us because we have to make decisions according to what we find when we first arrive at the shout.

Generally, as aircraft captains we have amassed enough experience to allow us to handle difficult circumstances, and find our way around them: we are constantly changing and adapting procedures so as to complete the task at hand successfully. We are never given inflexible direction; instead, we are usually presented with the opportunity to find our own way. The rule of thumb is, if you can make it work you do it, but if it's just too dangerous or not possible then come back and rethink it. But to save a life we will always do what we have to do.

Arguably one of the most courageous decisions to make is not to land on a IRT shout with a T1 casualty because it is too dark and there is too much blowing sand, knowing that the casualty may die, but the aircraft and the crew and the other passengers are too high a price to pay to save one life. The powers that be never drove us to make difficult and dangerous decisions that increased the risk to the aircraft and the crews. But it was the hardest decision to make, because every part of us was programmed to go in there and get them out, come hell or high water. The decision to abort a landing is one that is never taken lightly.

The IRT was made up of several elements: the Immediate Response Team itself, which included six medics, also known as the MERT (Medical Emergency Reaction Team), an EOD (Explosive Ordnance Disposal) team and the protection party, which was taken out of a 3 Para platoon, or the Royal Irish, who were wholly assigned to us at that time. They all did shifts of twenty-four hours and operated in the back of the cab, so it was jammed full of kit and people, usually being battered around by turbulence and 160 mph flight speeds at low level.

Prior to our arrival into theatre, the IRT had been working flat out taking CASEVAC cases out of the platoon houses on a regular basis. The HLS at Sangin was inside the wire and they had built a wall around the outside, which provided some cover. Now Zad's HLS was in a bit of clear ground, which made it easy to secure and land in.

The worst HLS, by far, was Musa Qala, and it was hideous. The Musa Qala District Centre had one possible HLS: it was outside the wire, and troops had to go outside to secure the site. Surrounded by high ground, it was a very tight and dusty landing spot.

The Taliban had started to lob mortars at the District Centres all day, every day, in the knowledge that eventually they would hit someone and one of the big fat fucking helicopters would come trundling over to pick them up, which they could shoot down. Brilliant, top plan, and quite frankly every day it was coming closer and closer to being a reality.

As the Taliban began monitoring the pattern of approaches into Musa Qala they began to realize that because of the limited number of options for getting in and out, the likelihood of taking down a helicopter was increasing, as long as they maintained their mortar assault. The worst part of it all was that we knew all this but there was very little we could do about it, because the guys that were entrenched in the platoon houses were hanging by a thread as a result of the constant mortar bombardment. What were we supposed to do, just leave them there?

Those guys were in there, making a political point, and war is a political business, but with this point came risk and effect, which in this case was that the guys on the ground from 3 Para were getting a thorough pasting. They were utterly helo-dependent and our cabs were their lifeline.

The lads were holed up in these shitholes. With the whole situation proving to be somewhat embryonic the DCs became outposts that could only be resupplied by helicopters. The Chinooks did swap troops in and out, but anywhere near Musa Qala, Sangin or Now Zad and it became a deliberate night op, unless it was a T1 like now. The lifts in and out were like a very dangerous crew change.

A message came into the Forward HQ via MIRC.

'*HQ RC South releases IRT Bastion. Call signs Doorman 26 and*

Wildman 50/51 to CASEVAC casualty serial 004457 effective 1100.'

The clock was on. Adam Watts and his crew spooled up to go.

'Mate, we need another door gunner,' said Watts.

'Take Sam because he's not seen Musa Qala,' I said.

'Can you please send a runner over to the IRT tent and tell Sergeant Hannant that he is jumping on this shout. Thanks, mate,' I said to a watchkeeper.

We stood back and let them get on with it. Adam and his crew briefed with the Apaches. Within minutes the guys had briefed then gathered their kit and legged it to the cab.

Once they had left the tent, Giles, Dave and I had an Int brief with J2 to update us on the general area in broad-brush terms.

The J2 Corporal began by discussing the situation in the various DCs in Musa Qala, Sangin and Now Zad.

'Just remind me, why are we holed up in these shitholes?' I asked.

The J2 chap explained: 'Tootal has had this situation forced upon him. It was meant to follow the Bosnian model of operations. 3 Para are in the villages because we need a presence there to provide security for the reconstruction. We set up the District Centres in order to support the villages and towns.'

'Aren't we just turning ourselves into fixed targets?'

'Yes, we are, but we have to deal with it because it isn't going to change.'

'Bloody hell, it's just like the Alamo.'

The Int guy nodded.

In JHF(A) there were four radios. One was a clear, or unencrypted, radio to the ground, which most of the time was kept available so that other ground call signs could use it to call into the HQ. There was a secure radio, which was inter-aircraft and worked back to the base with two air-to-air frequencies and two air-to-ground frequencies that were UHF. One, or both, of them would be used to talk to the ground controller for situational awareness (SA) and one was a dedicated radio for terminal controls.

I could hear over the secure radio the Apaches talking to the guys in the field. Terminal controls were used when the controller was trying to describe a target to the Apache remotely from the CP. There was often a lot of talking because of the complexity of the task. This would then dominate the channel so that nobody else could use it. This system meant that there was always one frequency where a controller could be guiding the Apache for a missile strike and another that was providing SA.

The net that was used by JHF(A) was also the same one that Apaches used to talk to one another. We could hear the dialogue from the battlefield being played out in the CP as the Apaches covered each other's tails. They were the link between the guys on the ground and JHF(A), as they sent back up-to-the-second situation reports.

Quite often the enemy mortars would stop firing when the Apaches were above the DCs, so it gave the guys on the ground some much-needed respite. The Apaches would fly around in slow lazy circles looking at things, scouting and patrolling the DC. Adam had reached the DC.

'Widow 75, this is Wildman.'

'Wildman, Widow 75 – go.'

'Wildman, two Apaches, standard load out, 300 by 30 Mike Mike, twenty-eight rockets and four Hellfire each. Fourteen miles to the south, to operate in the block 45 to 55. Playtime is 120 Mikes. We are escorting Doorman 26, he's twelve minutes out, ready for AO update.'

'Roger, here's the AO update. We are at the Musa Qala platoon house. We have taken fire through the night from the area of building 168. There is a lot of enemy movement in the area of spots blue three and four. We are concerned about being mortared now from spots red one and eight.'

'Roger, can you confirm that the LS is secure and cold?'

'Roger that – the LS is secure and cold.'

'Doorman 26, Doorman 26 – the area is ice.'

Two clicks from the Chinook.

'Widow, Doorman is ninety seconds out from the west.'

'Oh yeah, I've got him.'

There was a pause on the radio for a couple of minutes.

'Widow 75, Doorman 26 has lifted.'

'Widow 75, any areas of interest you would like to check?'

'Can you check out red spot one, mate?'

'Roger, stand by.'

'Widow 75, Wildman, enemy currently not apparent at red spot one.'

'Wildman – could you just give us another five minutes and look in the are—?' The voice petered out as the Apache cut in.

'Widow 75, I am sorry, mate. I have got to go, I am bingo fuel. Hey – we'll be back as soon as we can.'

'Yeah, all right. Y'know we always look forward to seeing ya.'

'Just shout, mate. Wildman out.'

These blokes are being shot at day and night so occasionally, once the Chinooks have landed on and have safely got underway again, the Apaches would hang around and use up the rest of their fuel to give them as long as possible without bombardment.

Adam Watts got back with the casualty. The T1 was a para who had been shot in the chest. He didn't make it. Our first night up in Bastion and a para was dead before we had even unpacked our bags. The next morning, we continued the handover. There was a resupply to Musa Qala so Dave and I jumped aboard the cab that Adam Watts had been using. Adam had moved up to HRF Captain.

I was sitting in the back as the rotors cranked up and the thunderous noise of the two powerful engines filled our eardrums with the usual sensory overload. We lifted to the hover and then slowly moved over to collect an underslung load, using the hook under the cab. As Adam attempted to lift the load, he found that the cab was too heavy. The aircraft started to reach its maximum PTIT – power turbine inlet temperature. This meant the engines were working at max temperature and hitting the upper normal limits – once the engines get to max temp they cannot produce any more power and either the engine will blow up or the speed of the blades, what we call NR, or rotor speed, will start to slow as a result of the lack of power to drive them round, and the cab will fall out of the sky.

The Chinook may seem like a phenomenal beast to most soldiers but it cannot do everything and sometimes the conditions or the weight of the load are stacked against us. The options here were to burn off some fuel or to reduce the overall weight of the aircraft. I was part of that overall weight.

One of the crewmen came up to me and signalled for me to get off. I disembarked and we left Dave on there as he was next up on IRT. Sam stayed on too. But it turned out that was my only chance to get into Musa Qala before the handover was completed. And that my first visit to the place would end up being rather more intense than a chance to familiarize myself with the lie of the land.

Chapter 12

27 Squadron had said that Helmand was a tricky playground as the Taliban had been attacking the DCs. They were putting RPGs through Musa Qala, Sangin and Now Zad. We quickly realized that the most challenging work was going to be part of the IRT.

They handed over to us on 2 September. The lads that we were supporting didn't have the measure of the new kids in town so it was always a delicate process to integrate with them in a way that didn't piss them off. Ultimately it was important to me that we got on and they rated us. They were not going to do that if we rushed in like a bull in a china shop so it was about treading softly, softly, in order to make sure I knew what was going on so that I could ensure I had their respect.

I eventually sorted the whiteboard and finalized a two-day rotation. Dave started the IRT, Giles held HRF and I was the spare and would pick up any other tasking. I didn't want to come over all bullish and give them the old 'I am in charge and therefore I am first in the hot seat' malarkey.

The rota system was working well. I came up on to the HRF line.

We had to go to Now Zad to drop off some mortars and ammo because the lads were out of them. We had to drop them off at the bottom of the hill. It was the middle of the night, pitch black, and there was going to be a JTAC (Joint Terminal Attack Controller, a soldier responsible for vectoring Fast Jet attacks on enemy targets) on top of the hill, using a laser only visible on NVGs to indicate the landing site. The whole crew was to be on goggles.

Daffers and I headed over to the cab. I had grabbed my issue life-saving combat jacket (LCJ), which contained the bare minimum of kit. It holds items like a torch and compass, a personal survival radio, flares, a heliograph, silk escape map (tough when wet), razor blades and windproof matches.

This was in case the aircraft was downed and we escaped. If that was all we had, then that was what we were trained to survive with. However, common sense dictates that any fool can be uncomfortable in a survival situation, so we tended to pack some extra kit. I carried a Camelbak water pouch in a thermal cover, half filled it with water and stuck it in a freezer. I slung it on the back of the seat in a cockpit where temperatures can approach 50 °C in the summer and enjoyed the luxury of cold water for a few hours in the blistering heat. If I chucked in a go bag containing a sleeping bag, roll map, extra ammunition and some warm kit there was a fighting chance that I'd make it through a night. It was a layer system; if we went down hard in a contact, I had what I was wearing, if I had a little more time I could grab my Camelbak, and if it was a slow-time survival situation in a relatively benign environment I could grab my go bag as well.

Desert combats and flying kit had improved since the Gulf War. We now wore a fire-retardant CS95 desert combat jacket underneath which we wore a long-sleeved shirt and combat trousers, with Nomex fire-retardant long johns and long-sleeved high-neck T-shirt. Even our boots were fire-retardant. I was always hotter than a snake's arse in a wagon rut and constantly sweating my cobs off when we added our LCJ and armour over the top, but it was better than burning to death if the cab caught fire.

Sam and Dan were already onboard when I got there. Daffers did the walk-round and we ran through the pre-start-up checks.

As we put on our goggles, Dan said over the intercom: 'There is nothing darker than the Afghan darkness.'

'Yeah,' said Sam, 'it always takes me a little by surprise even though I know there is no cultural lighting.'

'Right, gents,' I said, ' it's going to be quite challenging, getting this load down safely.'

It was the first time we'd done underslung loads at night into the dust. It was a completely different kettle of fish. It was pitch black. I thought: *This is fucking stupid.*

I knew that they were going to shine a beacon, to guide us in. Not visible to the human eye, the beam it emitted was picked up with night vision devices, including our NVGs. The Taliban did have some night vision devices but they were pretty old and basic so we continued to rely on the technique, i.e. the troops signalled from their pick-up location and we flew to the light – it could be seen from miles away on a clear night.

The drop-off was just beyond ANP Hill (Afghan National Police). We were coming up through the mountains, enveloped in darkness. The area was full to the gunwales with enemy wanting to shoot us down. We passed by a tight gully and the landscape opened out in front of us. We were tasked to drop 2 tonnes of underslung mortar ammunition in the dark.

I said to Dan on the intercom: 'As soon as it touches the ground you're manual release.'

I wanted to make sure that as soon as the ammo, which was contained in boxes in a net under the cab, touched the ground, it would instantly be released from the hook. I didn't want it to be dragged across the ground, potentially going bang under the aircraft. The best way to achieve this was by using the crewman's manual hook release lever and not the electric switches that we had on our cyclic stick. The manual release lever involved a piece of wire attached to the hook so there were no electrical circuits to go wrong. I had 2 tonnes of bombs underneath me, which I had to put down very gently so as not to set them off! In my head I was thinking: *Fucking hell, this is PhD-level flying. Our aim is to do this, not get shot at and not crash the aircraft in an area that I don't know and have never seen before. This is fucking ridiculous!*

Daffers, in my left-hand seat, was giving me good nav patter.

'You should be able to see the light on the nose now, twelve o'clock, two miles,' he said.

'Mate, I can't see fuck all,' I replied. 'Where the fuck is it?'

Then suddenly a long beam of light like the finger of God came into my field of view.

'Can you see it now?' he said.

'Yep, I have got the light,' I replied as I focused on bringing the aircraft down.

'They are shining it at the load drop point. Can you see that?' asked Daffers.

'Mate, I am going to take this one as it's our first one.'

Daffers and I shared the flying but as I was Captain on this mission I wanted to make sure that I had a handle on it. If anyone was going to fuck it up I wanted to make sure it was me.

'Yeah, fine,' Daffers replied.

I concentrated on the light piercing the darkness like the eye of Sauron. I set the aircraft up in the approach gate.

'One hundred feet, thirty knots,' called Daffers.

We were descending and it was fucking dusty. Dangling beneath the belly of the aircraft was a string of bombs in a net. And I was still plagued by the insanity of the situation:

Fucking hell, this is mental! This is absolutely crazy. What the fuck am I doing? It felt like I was going mad.

'Eighty feet, thirty knots,' called Daffers.

'Twenty-eight knots, coming down to fifty feet, twenty-five knots, twenty-five feet, twenty knots.' He continued to talk us down towards the light.

All I could think was: *I have never done this before in my frigging life and it's fucking hard!*

I was gently easing back on the stick. I was tense. I was really tense. The aircraft was coming down and I could hear a voice inside my head chanting: *Don't pork it up. This is fucking stupid. Why am I doing this?* Shooting tanks was a lot easier than dropping fucking explosive underslung loads.

'Ten feet below the load,' called Daffers.

He knew this because he had checked what height, on the RADALT, the load had come off the ground at Bastion and then added 10ft to it. It was a sign of a good co-pilot.

Sam had already picked up the patter as we hit 40ft. I was coming down on to the hover. On the gogs I could just see the horizon, followed by the dust cloud starting to billow up. As it was coming up into my field of view it was like being in the dunker and watching the water rising up through my goggles. The dust filled my NVGs' field of view until my vision was completely blurred.

Dan was in the hatch watching the load. Sam was scanning the dust cloud. Sam gave the cadence beneath the load. They couldn't see the ground and we couldn't see the load. Dan was trying to assess how far the load was off the ground but it was so dark there were no references – no texture, nothing to indicate the distance between the helo and the earth beneath other than the RADALT. It was pitch black, there was no moon, no shadow. It was like trying to look through a barrel of tar and estimate how far down it was to the bottom.

Dan noticed the strop that held the net to the hook went a bit slack, reached for the manual release handle and jettisoned the load. This guaranteed that the load would go. If we had opted to drop the load via the electric switches it could have been catastrophic. If one of them was to fail, we would still have had the load attached as we tried to climb away. This potentially could have caused the cab to rotate around the load, or worse drag the explosive load along the rocky ground. Under the circumstances, in the blanket of darkness that enveloped us, this would have been very bad indeed.

In my head it was frantic. I was thinking: *Fuck, I am going to lose all reference in a minute and then we are in a world of shit.* We were engulfed by a large billowing cloud of sand. Over the mike I sounded a bit panicked and called out: 'I am losing references. I am losing references.'

And then I heard 'Load's gone' from Dan.

The load dropped a couple of feet. Dan had released it early because he knew I was fighting the aircraft and conditions. It was safer to drop the load a couple of feet than for the cab to spear into the ground because I could see fuck all.

It was a real team effort to get that load down. I was flying it but Daffers was doing the non-handling duties, RADALT and all the other details to make sure that we didn't get into an unusual position.

Co-pilots follow through on the flying controls during tricky situations like this in case they may be able to see something that gives a reference to hover on when their pilot is blind. Sam had been voice marshalling the aircraft in the dust cloud while Dan had been head-in trying to monitor the height of the load over the ground. If Dan had thought it was too low we would have had a problem. We had aimed for, and achieved, load down, get rid, fly off.

The troops were out of ammunition and they had called for an EMER, an Emergency Resupply. It meant that we were flying in light levels that were below ones we would normally fly in. In order to deliver the load we had pushed the boundaries. We had operated above and beyond our normal training practices to give them a fighting chance. And we hadn't trained for it because it was well outside our peacetime limits.

All together, we breathed an enormous sigh of relief. I pulled pitch and we climbed to height and headed home. I could just sense that everyone in the crew was thinking what I was still thinking: *Fucking hell! We have not done shit like this before!*

I spoke to Sam. 'Did we drop it?' I tentatively asked him.

'A little bit.'

'How little's a little bit?' I cringed, shrugging my shoulders and pushing back into my seat.

'No, mate, it was only a couple of feet,' he reassured me.

'Did we wreck the load?'

'No, we didn't wreck the load. It just sort of tilted over a little bit.'

'Was it safe?'

'Yeah, it was safe.'

Relief took over. 'Ok. Good job!'

Flying on a dark night is hard.

Flying on a dark night and landing in a dust cloud is harder.

Flying on a dark night with an enemy who is trying to kill you and landing in a dust cloud is harder again.

Flying on a dark night and landing in a dust cloud with an enemy who is trying to kill you and with an underslung load is even harder.

Flying on a dark night, with an enemy who is trying to kill you, landing in a dust cloud, like thick talcum powder, with an underslung load full of volatile high explosives – that you cannot afford to drop hard on to the ground in case they go 'bang' – would be classed by normal people as total madness . . .

And as we headed back I realized once again the mammoth fucking task that was ahead of us. We were doing everything at night and I realized that nothing had trained us for this in the UK. Now we prepare because we have learnt from our experience but prior to that point I knew we all felt the same. We had never done this before in our lives.

This situation that we were facing had been generated by what was looking like a strategic error, which had been forced upon Colonel Tootal. It was due to the fixed positioning of the platoon houses. This meant we had to push boundaries to ensure the guys on the ground had the supplies that they needed.

The guys were stuck. They had no freedom of movement, which meant that their resupply was fixed, and only limited freedom of manoeuvre, which meant that we had to go to the same place and drop it off again. We therefore had to do resupply at night in order to mitigate the risk. You could train to some extent but you could never combine the landscape, the enemy and the circumstances, so when it became a live gig we were operating on a knife-edge.

We didn't know the area so we were operating blind. We were flying into the unexpected. It was stressful and tense because ultimately we had to complete the task at hand and return day after day to Bastion with our aircraft and crew intact.

As we came back to Bastion there was a feeling of elation among us all because we had done the job and we didn't fuck it up. Deep down we all knew that this was just the beginning of eight weeks of fucking hard work and that the operational landscape had changed. It was angry, it was dangerous and we needed to stay on our toes. The pressure was on and the snag was we had never operated in an arena like this before. Then the news came through on the BBC that there was an aircraft down.

Chapter 13

On 2 September we were in the IRT tent watching TV and the news came on informing us that some service personnel had been killed in Afghanistan. Some prat had speculated that it was a Chinook. On top of that the BBC news was reporting that the dead consisted of ten RAF, one Army and one Royal Marine. This was the exact make-up of 'B' Flight – my flight. We watched it in stunned silence.

Dan piped up first: 'That's just fucking brilliant.'

The loss of an aircraft turned out, tragically, to be real enough. An RAF Nimrod had gone down, but the supposed 'expert' the BBC had dug up had been listening to the Taliban military propaganda and he didn't have a fucking clue. The only thing that I knew for sure was that I definitely wasn't dead. I tried not to give it a further thought.

I decided to go and take a shower to cool off and freshen up. It was around lunchtime and Giles was walking back to the accommodation. He had been on the phone to his wife. As he passed me he said: 'You better phone your missus, mate, she's in a right state because of this crash. She thinks it's you because the news is saying it's a Chinook and there is one Bootneck confirmed dead.'

I was the only Royal Marine that flew the Chinooks and one of the few Royal Marines in theatre at the time.

I replied: 'Ok, I'll go and do that now.'

I knocked off the shower and went straight down to the phone cabin to phone my wife and let her know that it was all tickety-boo. As I got to the phones, Op Minimize was called

and all the welfare communications were shut down. Op Minimize is where all communication with the outside world is cut
off. It usually happens when there has been a casualty and they
are waiting to inform the next of kin to ensure only the correct message gets home. Oh, the irony. So I couldn't speak to
her until the next day.

Back at Odiham the DCOS had freaked out. Fortunately,
the patch community, non-official support network kicked
in. Digs, back at KAF, had managed to get a call into Ops
back at Odiham to say it wasn't one of our Chinooks.

I became slightly divorced from it. There was nothing we
could have done. We were trapped at Bastion thinking: *We
don't know who the hell it is but it's fucking none of us.*

What were the BBC thinking? They had completely jumped
the gun. The news game had become a dirty business with
24-hour news channels generating a greed for information
that too often came at the expense of truth. The BBC were
so impatient and they couldn't wait for the MOD to feed
them the information. They wanted to be the first to break
the aircraft type and so brought in their supposed expert. All
of the information pointed towards a Chinook because the
Nimrods were operating outside Afghanistan. It was an
unexpected turn of events.

The MOD have to inform the families first that their loved
ones have been killed. And in getting their facts wrong the
BBC demonstrated a complete lack of respect for the families of the serving personnel in theatre. Otherwise they would
have waited until the information was released.

The cost of this impatience was that the DCOS was sat in
our kitchen fretting, crying, worrying, watching and waiting
for the Station Commander and Padre to walk up our path to
inform her of my death.

I have lost friends in conflicts. I considered them brothers as they were Royal Marines. It's always hard losing Royal Marines but even harder losing mates that I knew well. I lost a close mate on the American CH-46 Sea Knight crash on the first night of Gulf War II. It was the assault on al-Faw, which was one of the first objectives of the allied campaign in Iraq: to take the oil industry in the al-Faw peninsula quickly and intact so that it couldn't be sabotaged or destroyed by the Iraqi military. This was a two-point plan, firstly, to prevent an ecological disaster – the flooding of the Northern Gulf with oil, as happened in the 1991 Gulf War – and secondly, to ensure a faster resumption of oil exports, which were vital to the reconstruction of post-war Iraq.

Following days of terrible weather, the assault on al-Faw was set for 20 March 2003, 2200 hours (local time). Prior to the operation US gunships and fighter-bombers bombarded the known enemy positions on the peninsula. At night in a typical airborne night assault, landed by helicopter, 40 Commando and US Marines captured over 200 prisoners. Concurrently, air and sea landings took the gas and oil platforms out to sea.

This was followed by a second assault by 42 Commando and USMC helicopters to land slightly to the north of the town of al-Faw in order to protect 40 Commando's northern flank. It was on this insert that the US CH-46 Sea Knight crashed, killing everyone onboard.

On the night of the crash I was onboard a Royal Navy Sea King Mk 7 ASaC (Air Surveillance and Control) helicopter from 849 NAS, in my capacity as the Operations Officer on 847. I was in the back of the aircraft to help interpret the ground scheme of manoeuvre.

The next night two of *Ark Royal*'s Sea King Mk 7 ASaCs, including the one which I had been on the night before, crashed into each other and all the guys that I had been with the previous night were gone. Three cabs in two nights. It was a lot to take in. I was stunned but despite this you have to get on with the job.

I have lost mates in America too. One was shot down in a Cobra in Iraq. I can understand it better when they die in battle. When people die in accidents it somehow seems like a more pointless waste of life. Moonpie died in an aircraft that had a double engine failure, Rosie died in an aircraft off the coast of California in a training trip and it was all absolutely pointless.

To some extent military aviators have a greater expectation of death. It goes with the job. I have felt more scared on my motorbike than I have ever felt in battle in a helicopter. It's not that I am invincible but I think you know when your number is not up. I don't think when I am about to get in a helicopter: *I am going to die today*, and if I did I would get straight out again. I do have a three strikes and you are out rule. If there are three things wrong with the aircraft then I think that someone is telling me not to go flying today and I won't go.

We are in a dangerous part of the military, which is aviation, and it is full of risk. Surprisingly though it tends not to give you a heightened sense of your own mortality, but that, as I was soon to discover, isn't always the case.

Chapter 14

At 0600 on the morning of 6 September it was just another day in the heart of the Afghan conflict.

There were briefs twice a day, one in the a.m. and one in the p.m. On the morning of the 6th the a.m. daily brief was run by Mike McGinty. Dan, Sam, Daffers and I had our breakfast as usual and headed over to JHF(A) CP for the morning brief around the bird table. The bird table was an enormous waist-high table, equivalent in size to three ping-pong tables pushed together. The duty Apache crews, the duty Chinook crews, the Ops officers, the groundies, the engineers and the Int guys all gathered round.

'Ok, let's start off,' asserted McGinty. 'The met for today, please.' And he looked across at the watchkeeper. The watch-keeper read from a piece of paper in his hand:

'Wind is 320 degrees at 3 knots, visibility 15 kilometres. Sky conditions clear. Temperature 27 °C. Dew point 18 °C. Relative humidity 10 per cent. Pressure 1021. The wind is due to pick up to 15 knots during the afternoon, with possibility of dust storms reducing visibility.'

'Int, please,' asked McGinty.

A fierce-looking, blonde lady Intelligence Officer steps up: 'Overnight there were troops in contact in Musa Qala, twenty-six rounds of 105mm were fired and one EKIA. There was a contact in Sangin, 3000 small-arms rounds were fired at target, seven Taliban killed.'

'Ok, aircraft serviceability, what have we got for today, please,' he said.

The chief engineers of both the Apaches and the Chinooks took it in turns to deliver their piece, updating us on how many aircraft were available and what flying hours available each aircraft had.

'We have some routine tasking this a.m. We need two loads to go, one to FOB Robinson and one to Now Zad. We'll launch both Chinook and AH together. Go via FOB Rob, do the drop, then the first Chinook can RTB and then we'll take the second Chinook and escort into Now Zad. And that's the lot. Thanks a lot, guys.'

The brief ended and Daffers and I headed back to the tent to brief up Sam and Dan. As we were the standby crew, Giles was IRT and Dave and his crew would pick up HRF. We completed the brief and headed over to the Ops room.

Back in the Ops room as soon as the brief was over and we had all been sent on our way, the MIRC chatroom system clicked into life.

'*TIC Garmsir.*'

The DC of Garmsir, in the south of the area, was a key strategic town which had fallen into the clutches of the Taliban. British commanders ordered that it must be retaken as a top priority. As 3 Para were locked in bloody battles in the platoon houses, it was left to a group of twelve British soldiers, including TA reservists and medics, to lead a force of Afghan soldiers and police, who were barely trained, across the Taliban-held desert.

'Call up the two spare IRT Apache crews,' shouted McGinty.

The two Apache IRTs' cabs were released by the Rear HQ and sent into Garmsir to offer some back-up to the troops on the ground. News soon filtered back into the Ops room that the contact was hot and the Apaches were engaged in heavy missile fire.

There was an Apache in the air that was liaising with McGinty, and the Apache and the Ops room were also speaking with the 3 Para JOC (Joint Operations Centre), located next door to JHF(A), via the secure radio. McGinty was getting the Int quicker than the JOC as the Apache pilot was hovering, looking out of the window and also speaking to the guy on the ground direct, who at the same time was speaking to his headquarters in the building at Kajaki, which was then relaying the Int back to the 3 Para JOC. Forward were receiving the Int faster than the 3 Para JOC. Even so the information was sparse.

Squelch: that radio was alive with the whooshing noise that sounded like a someone clearing their throat and slurping at the same time. A miked voice crackled in the background: 'Yep, keep left. Keep left. One. Zero. Alpha. Yes I can see some movement below.' Squelch.

McGinty was on the radio talking to the Apache hovering over the minefield site.

Squelch. 'One Zero One Zero Alpha, this is Zero Alpha, when you get chance send SITREP.' Squelch.

No sooner had he asked the question then the Apache cut in. He was talking to the guys in the field. Sending back a SITREP – a situational report – was not the priority and McGinty, a pilot himself, knew that the minute the Apache pilot had a chance to report back he would do so.

Squelch. 'One Zero Alpha, CONTACT . . . I think that was fine.' Squelch.

And then all of a sudden: 'One Zero Alpha, CONTACT.' The Apache's guns had opened fire thirteen to the dozen mid-chat.

Squelch. The aircraft had unleashed all its fury on the insurgents on the ground. McGinty finally called over the watchkeeper, who ran over and stood in front of him. McGinty held up his index finger to say 'Wait one minute' until he had finished on the radio.

Not long after this, the MIRC began to click once again.

'*3 Para casualties in Kajaki.*'

McGinty popped into the next-door tent to see Tootal. He got the nod to stand up the cabs. The room was a buzzing hive of activity.

The MIRC clicked.

'*Casualties Kajaki. Doorman 26 cleared.*'

'Right, gents, call in the Chinook IRT,' he shouted.

'Everything all right, mate?' I said to McGinty.

'We have casualties at Kajaki. We are stepping up to next door,' he replied.

'Do you want me to come through with you?'

'Yes, please, and can you sort the IRT.'

I asked the watchkeeper to get on the radio and call the IRT into the 3 Para CP. We moved next door and we were soon joined by Giles, who was the IRT Captain.

The picture was confusing. It appeared that a 3 Para patrol out of Kajaki had encountered casualties but we didn't know why.

'Sir, we have got three casualties,' interrupted a watchkeeper.

'From what?' said Tootal

'We are trying to establish that now, sir.'

'Where are the casualties?'

A grid reference was given and plotted on the bird table.

'That is pretty steep ground,' said Tootal looking at the topography.

Giles had a look and said: 'I am not sure we would be able to get the wheels on there, sir.'

'Sir, one of the casualties is T1.'

Giles and I were listening intently. We were waiting for Tootal to make a decision.

'What other assets have we got that could get in there?'

His Ops Officer replied: 'MH60 Pave Hawks' – a twin-engine medium-lift helicopter – 'from KAF, sir. They have a winch.'

'Contact RC South and see if they can be released,' said Tootal.

'That's going to take them forty minutes, sir, at best,' said Giles.

'That's too long,' said Tootal. 'Launch the IRT.'

Giles came over. 'I am going to go give it a go. If I can't make it, I'll route round to Lancaster.'

'Ok, mate, I'll see if I can talk to the guys on the deck.'

I got on the net and spoke to the guys on the ground to try and establish how steep it was. The reply was that it didn't really matter because they had a guy bleeding out. It was very tense because we didn't have the full picture of what was happening.

'Sir, update,' interrupted the watchkeeper

'Go ahead.'

'It appears the patrol has walked into a minefield. There may be more casualties.'

' We need another IRT,' said Tootal.

Slowly it emerged that there were more casualties on the line and the second IRT was stood up at very short notice. This was Dave's cab.

Dave and his crew were to route direct to Lancaster, one of the FOBs LS near Kajaki, in order to yomp in with a medical team to help the injured.

'Sir, update. The first IRT can't get in. It is also routing round to Lancaster.'

At this stage we had one IRT at Lancaster and the second one routing there as well. Now we had to wait for the medical team to walk the two miles from Kajaki into a live minefield to help the casualties.

'Sir, KAF have released the MH6os.'

It was now a race to see who was going to get to the casualties first, the MERT on foot or the MH6os. The clock was ticking.

MH6os arrived on the scene thirty agonizing minutes later and proceeded to ferry the casualties back to Lancaster, where they could be transferred on to the waiting Chinooks and evacuated back to the field hospital at Bastion.

As time progressed I thought about crew duty for Giles and Dave. We were going to have to swap and my crew would have to stand up to IRT. Giles and Dave returned around 1500. I agreed with Dave that we would be standing up.

I headed over to the IRT tent to let the crew know and to brief them. I walked into the tent. Sam and Dan were sat playing on the PS2.

'All right, lads, we have been stepped up to IRT,' I said.

They looked up, nodded and then got back to their game. Daffers was lying on his cot reading.

'It was going off in there, mate. There was a mine strike at Kajaki. A T1 and three T2s and two of the AHs are pumping out the rounds in Garmsir,' I said.

I kicked off my boots and sat on my cot. Dan and Sam stopped playing the PS2 and Dan looked up.

'A bit of *Firefly*, gents?' suggested Dan.

'Hoofing! I think that is just the ticket,' I said.

We needed to be on crew duty, available to work for sixteen hours in a 24-hour period. Quite often from ten o'clock there was nothing to do so you slept until 7 a.m. and then you had had nine hours' sleep. If you had a quiet night, you got a lot of sleep; if you didn't, by 8 a.m. you would be hanging out of your hoop.

If you had a quiet night, it was ok as long as you got straight up and got changed so that you didn't get a shout in a towel.

A pound to a pinch of shit, if you had quiet night you were going to get a shout first thing when you were in the shower. So we never showered at the same time. We always made it so that one front end and one rear end (one pilot and one crew) were abluting and showering simultaneously. If something happened then one of the crews could sort the stuff out while the other finished dressing. Me and Sam were together so that we could tab as part of our ablutions and Dan and Daffers were together. A cigarette and can of Coke set me up for the day. If I smoked it kept me on edge a bit. We can't drink booze so a caffeine and sugar-based chemical concoction was the next-best fix.

Chapter 15

As things had quietened down after the Kajaki incident I was lying on the rack, watching *Firefly* on DVD with Daffers, Dan and Sam.

The American green frame cot I was lying on was not the height of luxury. I had two combat pillows propping me up, and I was stretched out. I could feel the hard frame pushing into my back as I put my hands behind my head and crossed my legs. In the background there was a constant hum from the air conditioning. The smell of 'eau de bloke' hung in the room. It's a combination of dust, feet, sweat, fart and stale coffee. There were eight cots in total, four on either side of the tent, and the TV was in the corner. As you walked in the entrance there were shelves storing the filter coffee, porn, or 'Frankie' ('Frankie Vaughan') as we liked to call it, and weapons-cleaning kit. Everything a bloke needed to while the hours away.

We were now waiting for the call. The first IRT shout would not be my first time into battle but it would be my first time into battle in a Chinook, an aircraft that was now central to the coalition effort in Afghanistan.

UK armed forces really struggled to find an appropriate role where the full capability of the aircraft could be exploited. It was mainly being used to evacuate casualties until immediately after Gulf War I, when between July and October 1991 the Chinooks and Sea Kings were onboard a ship coming home and the humanitarian crisis in Turkey kicked off.

The Chinook force was then deployed to Turkey to help

with the distribution of humanitarian aid, rations and building supplies. In Turkey, where the aircraft were based was virtually unreachable by vehicle but there were thousands and thousands of people in need of food and protection from the oncoming winter. The terrain in Turkey, in the east towards the Iraq and Iran borders, is unbelievably hostile. There are massive mountains rising to between 10,000 and 20,000 feet, before they plummet down into cavernous ravines with no roads or other infrastructure, and no crossing points between east and west other than for goats and peasant farmers.

This marked the CH-47's introduction to its role in distributing humanitarian aid and relief in remote and difficult places, as it went on to do during the catastrophic earthquake in Pakistan in 1992 and as it still does today.

A lot of the work that we were doing in Afghanistan was in remote terrain that is difficult to cross in land vehicles, and in high regions such as the upper foothills of the Hindu Kush, where there are no roads. All the skills being put into practice in such places were acquired from the experiences of UK military Chinook pilots in Turkey in 1991. Skills such as landing on a knife-edge ridge with just the back wheels on it were developed and applied there.

This paid dividends in 2002, when the Chinook force first entered the combat theatre in Afghanistan and had to fly in the high-altitude terrain surrounding Kabul in technically very demanding circumstances. Each individual aircrew become very independent as every day they were forced to deal with the new, demanding and dangerous conditions that were continually thrown at them.

As pilots and captains we are expected to be able to act independently, confidently and slightly remote from the system of supervision. Otherwise the aircraft can't be exploited properly. The success of the Chinook squadrons is down to the experience of the captains, experience gleaned from the variety of roles we have played in different parts of the world. It's impossible to prepare for everything but we are prepared for the unexpected, which was just as well.

Chapter 16

'So it was pretty nasty up in Kajaki this arvo?' Daffers said.

'Yeah, I was in the Control Post at the time trying to get the guys on the deck. Let's hope it's quietened down now we are on watch,' I replied.

I spoke too soon. No sooner had the words left my mouth when the voice of the JHF(A) CP's watchkeeper called us up on the radio. There were casualties in Sangin. We gathered our shit and pegged it out to the cab, starting the engines with a thunderous roar. Some twenty minutes later we were airborne.

We had lifted the CH-47 and we were headed to pick up the casualties. I was pumped with adrenaline as we gathered speed, routed north out of Bastion, headed for Sangin and climbed to height. The crewmen looked forward to sitting on the ramp test-firing the guns. The guns were fired at height. There was a linear feature called the deconfliction line. North of the deconfliction the M60s could be tested safely.

'Checking weapons,' said Sam over the intercom.

'Checking weapons,' said Dan.

'Go ahead.'

Dan fired first and I could barely hear what sounded like a painfully slow rate of fire from the Pig. The reality was that the gas-operated M60 machine gun was chucking out 550 rounds a minute. The Minigun is an awesome weapon capable of firing up to 4000 rounds of 7.62mm bullets per minute. Sam then check fired the Minigun. The cacophonous blast of six barrels of electrically powered thunderous flatulence drowned out every other sound of the aircraft, not nice to be on the receiving end of but great for us to have defending the aircraft.

The smell of cordite filled the cockpit. I love that smell. It always reminded me of a toast that we would make when I was flying Cobras. With glasses raised, we would say: 'Gentlemen, gunpowder and pussy. You live for one, you die for the other, but you love the smell of both.' The noise of the Pig, raining down bullets, was just a distant whirr against the cascade of noise caused by the rotors.

We flew north across unpopulated desert, a flat, barren wasteland known as the GAFA – the Great Afghan Fuck All – towards the Sangin DC.

Here be dragons, I thought. We only went in if we had to. Either resupply or IRT, but it was so dangerous that the resupply had to be approached and planned as a deliberate operation. On IRT we took our chances, crossed everything and prayed it wasn't our time. This was one of those moments.

The British public had just seen the pictures in the papers of the Paras in the press defending the walls at Sangin in just their underpants. They were getting their arses handed to them by the Taliban and taking a lot of casualties.

At about ten minutes before we were due to land I wanted to make sure the crew were happy.

'Right, Sam, Dan. This is the plan. We are going into Sangin to pick up a T1 and four T2s. We are going to come in at height and then drop down into the gate. We want to come in hard and fast. Ramp down, casualties on and then get the fuck out of there as quick as possible. Are we good with that?' I said.

'Roger that,' they both said.

'From now on I am Mark, this is Daffers, you are Sam, you are Dan. Are we clear?' I said.

We weren't going to go with that 'sir, number one crewman, number two' bollocks. It was first names only on my cab. This little chat was my way of getting them ready to go in. We did it every time before an approach. They were used to it and this procedure was a way of getting everyone's head in the zone.

'Right, is everyone happy with what we have to do?'

'Yep,' Daffers, Sam and Dan answered in unison.

I then let them know that we were beginning our descent.

'Descending!' I said over the intercom.

We began the tactical descent, which is as close to falling out of the sky as is possible. It's with very little power but manoeuvring the cab in as many ways as possible, which makes our approach very unpredictable. It's harder to aim at us if we throw it around as we head towards the landing site. As I held the big helicopter in a spiralling descent towards the ground I knew it was all about to kick off. This was the last safe moment and every one of us onboard knew it. Senses were heightened and thoughts were focused. We were calm and collected but ready.

The Chinooks weren't equipped with secure radios. On the one hand this was a good thing because we didn't get distracted by unnecessary airwave twitter, but in other respects it was bad because it meant we didn't know what was going on and relied on the Apaches to relay information to us.

The Apache scooted forward to have a look at the area. The mere presence of the Apaches often inadvertently notified the Taliban that a Chinook was inbound. Procedures were such that the arrival of an Apache upfront was an indication that a Chinook was to follow shortly afterwards. There was nothing we could do about this and we weren't going anywhere without the Apache cover. But it was partly why the risk of being shot down was amplified. There were very few ways we could mix up our approaches because of the situation of the LS.

The aircraft moved into a holding pattern.

In order to get us into Sangin, the troops on the ground had to put down suppressive fire, 360 degrees around the base. Unfortunately, news of this hadn't reached our ears. Tracer cascaded from the walls surrounding the base and took us completely by surprise. We were two minutes out.

'Tracer coming out of the Sangin compound,' shouted Dan.

In the back Dan and Sam were poised on the guns ready to unleash fury if necessary.

'It's not clear where they are firing. Hang on, I can see tracer rounds coming back from the valley,' he said.

'Clear at the back,' said Sam.

Daffers and I looked at each other.

'What the fuck!' we said simultaneously.

'I guess we are going in there then,' said Daffers glibly. As we descended he began the talk-down into the centre of the tracer fire.

Sam was on the ramp and he could see little dots of Taliban on their bikes reaching for their mobile phones and informing their brothers that there was a Chinook inbound.

We would be flying along and we'd think we could just let these people have it. But we would always hesitate and not fire. Firstly, because we would have to live with ourselves, and, secondly, because we never knew for sure whether the bloke on a motorbike, on his phone, in the desert, was phoning his Taliban insurgent buddies and saying: 'They are on their way in and they are coming in from the west'; or: 'Hiya, Mum, I'll be home in twenty minutes for me tea. You can warm up the naan bread.' And if he was on the phone to his mum, and suddenly a Chinook opened fire on him, then his life was taken for no reason. But, on the other hand, if he was telling his mates to standby with the RPGs, then really we should have wasted him. This was the dilemma we faced.

It was dusk and the red and green tracer was illuminating the twilight. It was like heading towards a battle in a scene in *Star Wars*. I was so focused on the landing that I couldn't mentally absorb the reality of the rounds that were being fired beneath me. I had never flown into the middle of heavy fire before. The fading light exacerbated the colour and brightness of the tracer. It didn't seem real. More like we were starring in our own war film.

It was hard for my brain to interpret this situation in any other way. I was bizarrely hypnotized by the unusualness of it but we were coming in so fast and I had so much to focus on that I couldn't think that we could get hit as well. My head just didn't consider that eventuality. I was looking at the rounds being fired beneath the cockpit and the plumes of smoke rising from the ground but I couldn't see them. I was in the zone.

Daffers was talking to me constantly, a stream of information about bearings and distance to the target, counting down our height above the ground as we charged towards it. *We're going in . . .*

The Taliban guns were firing constantly. People often forget that the Taliban were brave enough to stay around to fight, which was why we ended up with so much shit being thrown at us. The Taliban wanted to get a Chinook without a shadow of a doubt – Intelligence had reiterated this – and they were definitely prepared to have go. They were relentless in their pursuit of this task and that was one of the reasons why the troops on the ground were fighting so hard.

Daffers continued to count down. And all I could think about was the landing. But I felt good. On top of my game as the aircraft responded to every instinctive adjustment from me. But the landing was what it was all about.

I don't want brown-out, I don't want to rip the wheels off, where's the casualty? I want to be in and out fast, bring your speed under control, bring it back. I was talking to myself.

As we approached from the west I hauled the cab into a right-hand turn. Pulling on the collective and cyclic controls, I flared the big Chinook to reduce the airspeed and control the descent. Then I levelled the nose. We were in the gate for the approach.

Daffers now changed his chant to height and airspeed.

'Fifty feet, twenty-five knots in the gate,' he confirmed.

'Fucking hell. It's going off,' said Dan at the door.

The RADALT bug made a slight 'whoop' and I cancelled it. This little noise told Dan that the aircraft was 50ft above the ground.

'Clear below down, forty,' Dan said over the intercom.

Sam started looking for the dust cloud.

'Dust cloud building,' said Sam.

'Thirty!' said Dan. 'Twenty!'

'At the ramp,' said Sam

'Fifteen,' said Dan

'At the centre,' said Sam

'Ten, eight, five,' said Dan.

'Front door,' said Sam.

'Four, three, two, one, WHEELS ON!' said Dan firmly.

'With YOU!' said Sam, almost simultaneously.

I glanced down through the glass of the cockpit to make sure that I wasn't too high to bring it down on the wheels. Then the aircraft touched down and the world disappeared into a huge cloud of swirling sand and grit. As we landed I relaxed my grip for a second. It didn't last. The sharp, sulphurous smell of cordite mixed with the pungent odour of hydraulic fluid filled my nostrils.

'Ramp down,' said Sam.

The ramp moved under its own weight, even though it had hydraulic actuators.

In the cockpit, looking out from beneath the shade of the spinning rotors at the unstoppable path of supersonic tracer arcing across the sky, I felt like a sitting duck. We were exposed and vulnerable, a very large, obvious target just waiting to be blown up. I didn't like it at all and was keen to get back into the sky and claw back a little more influence on my fate.

I was nervous. As my fingers anxiously drummed out rhythms on my legs my mind turned over the same two words again and again: *Come on, come on.*

Sam kicked stuff off as quickly as possible. Medical supplies, food, water, but he literally chucked it off. It was for the guys on the ground to rearrange once we were out of there. We didn't want to be on the ground for any longer than twenty seconds. As Sam was kicking off the supplies the troops on the ground were bringing on the casualties.

Sam stepped off the ramp. The world slowed down as he watched the bedlam around him. There were troops up at the walls firing. There were rounds coming in. The onboard protection team had run forward and were firing. The lads on the wooden walls outside the LS were shooting down. The twenty seconds on the ground felt like an hour.

The troops crouching nearby waiting to get the casualties on poured into the cab. Dan was focused on getting everyone onboard who was meant to be onboard and no unnecessary passengers. During the handover Sam had gone into Musa Qala on an IRT shout and they had inadvertently brought back to Bastion one of the on-the-ground doctors. This meant that Musa Qala didn't have a doctor and the doc then had to be taken back by another cab. Every time a cab went into one of these landing zones the odds on being shot down increased. It was important to avoid wherever possible any unnecessary trips into the hot spots.

Amidst the carnage the crewmen had eyes on who was getting on and off, trying to make a mental tally. The on-ground troops brought up the T1 on a stretcher. Sam took his harness off, walked out and helped on a casualty who had been shot in the leg, and then two more were brought on stretchers behind them. In the background the crewmen could hear the thumping of the .50 cal pounding the area. It was so loud that it could be heard over the reverberation of the turning rotors.

The casualties thought of the helo as a safe haven because it provided them with a passage out of hell. They relaxed once they stepped on to the ramp, lost some of their alertness so hung around a bit. Sam and Dan were edgy. For them the threat was far from over. They knew they were not safe yet so they hurried the casualties along.

'Get on! Get on! Get on!' shouted Dan.

The longer we stayed on the ground the more time we gave the Taliban to fix the target in their sights. The helicopter was a big, steel hundred-foot target full of fuel. We were basically a huge flying, rumbling bomb and if we got hit in the right place we went 'BOOM'. The casualties weren't safe until they were in the hospital in Bastion. That's not a conversation we would ever have with them though.

Sam tripped coming back on to the cab and dropped a .50 cal barrel on to one of the casualties.

'Sorry, mate,' he said.

He looked up and there was a large 'M' daubed on to the man's face. The casualty rolled his eyes and put his head down again. He was too out of it to notice.

Just at the last minute, before the ramp was lifted, a soldier ran up to Dan and chucked a sack of mail at him. In the midst of all hell breaking loose and getting the casualties onboard, this mail sack was really important. Mail was the connection between the guys holed up in shit and their families; the troops reaching out and their families touching back in. We took the mail as often as possible. Sometimes it was the smallest things which boosted morale in the biggest ways.

Dan provided the commentary. 'I have one on, two on, five on.'

All of a sudden the Apache came over the radio.

'Doorman 26, this is Wildman – you may want to hold off,' he said.

'Too late, mate, I am already on the ground,' I replied.

'You need to expedite, enemy setting up to the north,' said the Apache. Then the tearing sound of high-velocity automatic cannon fire ripped across the airwaves.

There were multiple targets at Sangin and I could hear over the radio that the Apache was already engaged in heavy fire elsewhere.

'Right, lads. Get clear, we have enemy setting up to the north,' I said.

'Which way is north,' said Dan, mentally prepared to aim the gun out of the correct door.

'It's in the six o'clock,' I said, so he now knew where to look.

As Sam took his final steps on to the cab, he looked back and could see pockets of dust popping in the sand as bullets rained down towards him.

'I am getting in the cab now. Just going to sit in 2 tonnes of fuel while being shot at,' he said sarcastically as he started lifting the ramp. 'Ramp up.'

'Lifting,' I said on the intercom and pulled pitch.

'Clear above and behind.'

We were good to go and I wasted no time.

'Pulling pitch,' I responded as I hauled on the flying controls and 17 tonnes of shaking, vibrating metal ascended into the sky. Once again we were engulfed in a cloud of dust and instant brown-out. And that wasn't our only problem.

As my left hand pulled up on the lever, I could hear the noise of the engines increasing rapidly as they responded to my request for power, and lots of it. The gearboxes began to whine as the big blades battered the air into submission, trying to lift the heavy aircraft off the ground and into the air as quickly as possible.

'NR is good with two good engines coming up together, good rate of climb,' said Daffers as we lifted up into the air.

'Above the light, above the noise' came the next call from my co-pilot.

This meant that I as soon as I could see we were clear of the dust cloud created by our rotor downwash, I could stuff the nose down and accelerate hard to get away from the HLS and back towards the life-saving care our passengers required at Bastion.

As we lifted, the T1 casualty was in a bad way.

Dan moved from the left side to the right side of the cab, constantly checking to make sure we were clear. Over his shoulder he saw a stitch of green tracer. It was big, so it was something quite hefty coming across and it tracked the cab.

'Tracer, BREAK RIGHT!' said Dan.

In a nanosecond I pulled in power and snapped the aircraft over to the right. The cab lunged away as we danced around the skies like the pink elephants on parade, dodging the bullets that were being fired from the treeline below the LS.

'We're taking fire, rear right,' the crewman shouted to me.

I pulled in loads of pitch, stuffed the nose down and rolled right to come over the top of it low and fast to the west. I snapped the cyclic to the right and rolled the big helo hard over the top of a line of tracer. As we manoeuvred violently away from it I saw the heavy red light of the tracer zipping below us from my eleven o'clock. *Shit*, I thought, *good call, Dan, if we'd come off right a split-second later it would have fucking hit us.*

As we clawed our way up through the thick, swirling dust, more big, fat tracer rounds whipped past above the aircraft. The crewmen were completely exposed, but returned fire immediately with their M60s. This at least alerted the guys on the ground that we were taking rounds. And within a few seconds I watched as the full force of British firepower poured down on the Taliban position.

As we climbed away I spoke up over the intercom. For all the drama of our departure, I had to remember we had badly wounded casualties onboard. I gave the onboard surgical team the option: 'Fast and rough or slow and smooth.'

'Bloody fast and who cares about rough,' was the reply. One of the guys was bleeding out and we needed to get him back fast.

Once we hit a safe altitude, the cab became quiet apart from the distant throb of the rotors.

'Is everyone ok?' I said over the intercom.

Everyone took a moment and checked themselves and then checked the aircraft. Sam looked at the maintenance panel at the rear of the cab.

'We are ok. I don't think we got hit,' said Sam.

While I urged the Chinook forward, the MERT were trying to stabilize the T1. Incredible, I thought. They were conducting ER surgical procedures in the back of Chinook while bouncing around like a ragdolls in a washing machine.

Dan pulled the curtain across to shield us from what was going on so that we could stay focused on getting the aircraft back to Bastion safely.

'How's he doing?' I asked.

'Errr . . . not so great,' said Dan.

The crewmen wanted him to be saved. We all wanted him to be saved. However, the horror of what they witnessed would be etched into their minds for ever. They watched and hoped that the T1 would make it. It was a balance between looking after the guys in the back, watching the MERT operate and keeping the cab in check. If they didn't pay some attention to him then they felt that they would be denying the casualty his humanity, but at the same time they couldn't just stand there and morbidly stare at what was happening. They would always wonder if they had done enough for him.

Down the back Sam and Dan were covering our rear and getting constant updates from the docs. Nose dipped, we routed home as fast as we could, red-lined at 160 knots, the Chinook's VNE – or 'velocity never exceed' speed. Not that rules and regulations were ever really a consideration that came into my head; it was as fast as the Chinook was going to go without tearing itself to pieces. I had never flown one as fast, but the only priority was to try to get the casualty back alive. Yet the Chinook now felt desperately slow. The desert whizzed past underneath, but time seemed to stand still. Twenty minutes had never felt longer.

'How's he doing?' I asked.

'We need to get back quick,' replied the doc.

I knew it. Over the intercom I could hear the doc talking as they worked.

'He's stabilizing. There's not much time but there's a chance we can save him. Hold him. Steady. Steady,' he said.

As we came in on the approach I could see the Land Rovers by the hospital HLS and, without sacrificing speed, I took the approach and landing as softly as I could.

As the cab settled heavily on to the main gear, the crewman

spoke over the intercom: 'Ramp down. Ambulance approaching from the rear. CO 3 Para and the RSM are here.'

As the casualties were offloaded, I turned and craned over my left shoulder, looking back towards the cargo hold. The Chinook crew, still pumped with adrenaline, looked back. I met their eyes, which were wide open in the midst of their grime-covered faces.

'Good job, lads,' I told them. 'We got 'em back.'

I felt an enormous sense of relief before Sam announced: 'The docs are coming back onboard.'

I watched from the cockpit, the rotors still turning above my head as, wearily, the doc put his headset on and spoke into the mike: 'Sorry, gents,' he told us, 'he died on the ramp.'

Fuck. I slammed the metal frame of the cockpit glass with a balled fist. *Fuck*.

'Fuck!' I shouted over the intercom, and slumped back in my seat.

Chapter 17

It was merely minutes after the doc had told us that the guy had died on the ramp when we got a call on the radio from the CP.

'Doorman 26, this is Zero Alpha.'

'Doorman 26 send over.'

'Zero Alpha, reposition for refuel, stay rotors turning. Out.'

'Right, gents. Looks like we have another shout.'

I pulled up on the collective, lifted to the hover and then moved slowly over the Hesco barriers that divide the HLSs to one of the refuel points. The crewmen had a big tidying-up job to do. It was bad enough that we had lost the T1 but the crewmen were now cleaning out the cab with a large roll of blue paper called Kim Wipe. There was no glamour in the job, and in addition no hoses or running water of any sort, just blood, sand, dust and elbow grease.

While Dan was cleaning out the back of the cab, Sam disconnected from the intercom and began to refuel. He placed the hose into the central pressurized fuelling panel and then began pumping just under 3 tonnes of gas into the aircraft. The CH-47 had two methods of refuel: pressure refuelling, which filled the tanks from empty in under ten minutes; and gravity refuel, just like how you would refuel your car, which takes an awful lot longer. Generally we prefer to pressure-refuel where possible.

We were having a chat over the intercom. I asked Dan: 'What was he like?'

'Yeah, yeah, he was in a bad way,' Dan said. 'They did their bit but . . .' His voice tailed off.

There was a moment of quiet. I then asked: 'Is everyone ok? . . . Daffers?'

'Yeah, I'm fine,' he replied.

I briefly paused. 'Dan?'

'Yep, I'm good,' he said.

'How's Sam doing?' I asked.

Dan looked up. Although we were pumping 3 tonnes of fuel into the cab at high pressure, he could see Sam pacing up and down near the fuelling area smoking hard on a cigar. We have a rule that states no drinking up to ten hours prior to flying and no smoking within 5oft of an aircraft. Sometimes we reverse this to no drinking within 5oft and no smoking for ten hours. Perhaps Sam had interpreted this quite loosely: at least he had moved a token distance away from the fuel . . .

'Yeeeeeeeaaahh . . . err, he's ok,' he said.

Sam had definitely had better days. To be honest I think we all had. We took losing a lad personally. All the 'what ifs' were flying through our heads. What if we had got there earlier? What if we had been quicker to start? Could we have saved him? As I sat in the cockpit, the thought nagged away at me.

The lads on the ground think that Chinooks will just come in and rescue them. If it was up to us we would come every time as soon as we were needed. But what the lads sometimes don't realize is that when they need a Chinook someone higher up the food chain is going through some risk-mitigating equations, weighing up the pros and cons, rolling dice, looking at chicken bones and then after some hours of deliberation responding in either the affirmative or the negative. By which time the casualties are bleeding out and essential time has been wasted. The Chinook crews would go anywhere, any time, to save the lads. If we were not sent it was *never* our call. We were there for the lads 24/7. The decision to hold us back was made by senior officers, not the pilots and the crews.

Some of the RAF guys were pushing boundaries that they never could have anticipated encountering. Matt Carter was a real hero. He was awarded the Military Cross that he richly deserved for incredible bravery and courage.

He was deployed with 16 Assault Brigade on an operation against a suspected Taliban compound outside the town of Now Zad in Helmand Province. During the first of three contacts, he co-ordinated and directed close and accurate attack helicopter fire support with devastating results for Taliban ground troops. During the last contact he left his vehicle, fearlessly exposing himself to significant risk as he forced his way to the front of the firefight to join the forward troops. This enabled him to direct aerial cannon fire against a determined enemy 30m in front of him. This risk was essential given the ferocious weight of the incoming fire from the Taliban.

His direction of these engagements proved critical, destroying the enemy location completely on one occasion. He remained with the lead dismounted elements of Patrols Platoon and took part in the immediate compound clearance. During this time Carter repeatedly exposed himself to a significant chance of being killed, and because of his gallant behaviour in supporting his unit he enabled the Patrols Platoon to regain the initiative.

Later on in the detachment, he participated in a battle group operation to capture or kill a high-value Taliban leader. During the insertion to the helicopter landing site the first wave of Chinook helicopters were heavily engaged by Taliban machine-gun and RPG fire, causing the aircraft to lift off again after only twenty seconds on the ground. Fearing being left behind on the aircraft, Carter jumped some 15ft from the tail ramp into the darkness, realizing the vital role he had to play in calling in air support to suppress the enemy. Immediately he got into the cover of a nearby ditch and called in an aircraft to destroy the principal threat of an enemy machine gun. He controlled the aircraft's heavy attacks, which destroyed the Taliban position only a few metres away from his own location, an action that proved to be decisive and allowed the remaining aircraft to land the rest of the battle group to complete the mission successfully.

Matt was part of the RAF Regiment. He was trained in soldiering but the crewmen were not. Nothing could have prepared the crewmen for this level of exposure to combat. They were aviators who were trained to be an integral part of a Chinook crew. They were not soldiers and at this time none of us were prepared for the Afghanistan we had encountered, but least of all them. We had come into theatre expecting to be peacekeeping and helping to rebuild communities – the level of opposition was unexpected. They watched helplessly as people's lives hung by a thread in the bloodiest of circumstances.

A lot was asked of the crewmen. Their primary duties were to manage the aircraft rear of the cockpit. One of their secondary duties was to fire the weapons. Inevitably this meant that they were likely to take someone's life with these guns. To take another human's life is the biggest thing a person can be asked to do. While it was foreseeable that one day this might be asked of them it didn't make it any easier for them to accept or live with.

Soldiers are trained to be desensitized to killing. They are trained in hand-to-hand combat, they fire shots at targets and bayonet dummies in preparation for their war. It doesn't make it any easier when it is asked of them but it is more expected that they will one day kill someone. It was being asked more and more of the crewmen, and this often took its toll. The Chinook crewmen are remarkable and do a brilliant job. I feel honoured to count them among my friends.

We were turning and burning. Once our tanks were topped up we disconnected the fuel hose, and closed up all the panels. I nudged the power and hover-taxied back over to the passenger HLS. Once we had repositioned ourselves on the HLS, Sam said over the intercom: 'Mark, we have got somebody at the ramp.'

Dave Westly popped his head into the doorwell and said: 'Mate, we have another one, a T1 at Musa Qala. I am your escort. The ground commander wants you to come in from the north-east.'

We had never approached this way before and it would add an element of surprise.

'Right, mate, it's all SOP,' I said – standard operating procedure. 'I want a two minutes call and I'll let you know when I am letting down.'

The brief was quick. Quicker than we would have accepted under normal circumstances but as it was a T1 we needed to get a shift on. Plus it was my old mucker Dave and we had flown together for years. I was confident that we had it all under control.

The light had gone altogether. It was now pitch black. Afghanistan has no cultural lighting. We pulled out the NVGs. We were all pumped from the day's events. The adrenaline had kicked in. This was the job and we always wanted to do the job well.

Out of all the crew that night the only one of us that had been into Musa Qala before was Sam, prior to the handover with 27 Squadron. Having had such success with the light beacon on the mortar drop at Now Zad and because none of us other than Sam had been into Musa Qala, we asked Dave to relay a message forward.

'Dave, as we come into the LS can you ask the JTAC to arrange for the guys to signal clearly so we can see it when we come in to land?'

'Will do.'

With that he disconnected from the intercom and headed off to start up the Apache, while we prepared to lift.

Musa Qala was a real shithole to find. It was particularly difficult because it was such a grotty little place. The troops in the DC had Taliban digging into the walls two feet away from them and trying to shoot at them. They were a target for everybody. The HLS was no bigger than 200 square metres, which was indicated by a small square on the map.

Us Brits in agreement with the town elders had established a 200m exclusion zone around the DC. We had agreed with the elders that anyone within the exclusion zone who shouldn't be there was fair game and effectively could be shot at.

Now, obviously we weren't going to begin shooting at people indiscriminately but it was a way of establishing boundaries. The town elders pragmatically agreed to the exclusion zone and kept it clear of civilians. The LS was in the middle of the DC, which was in turn surrounded by the exclusion zone.

In preparation for the approach into Musa Qala, Daffers checked the map for significant features to identify the LS. It was important that we had a good clear idea of where we were going into because we couldn't afford to have our heads down looking at maps trying to land on while the enemy threw ten barrels of shit at us.

Daffers was really struggling with the satellite imagery and LS photos because Musa Qala had had the crap kicked out of it and didn't look like much.

'Mate, check out these images. The site is there, that mud

hut is there, but that feature no longer exists. This is going to be a pig to go in,' said Daffers encouragingly.

We were all tense. We hadn't even left the cab, the turn-around was so tight. We had been shot at at Sangin and had just lost a bloke in the back. Although I had never been into Musa Qala its reputation went before it. The guys on 27 had told me what a nightmare getting in and out had been.

I looked at Daffers. 'That HLS is tighter than a gnat's arse-hole.'

We were varying the approach by going in via the north-east but at the end of the day we were going into the same landing site.

We all took our positions. I settled myself back into the right-hand seat. Over the intercom I briefed the guys in the rear: 'Lads, we are going in the long way around and will cut across the upper Sangin Valley,' I said.

'Roger.'

Once again all we could clearly hear were the exchanges over the intercom and radio. It was like entering another dimension – the aviation zone. This separation between the cockpit and the outside world kept us focused on the job at hand. It was already turning out to be one of those days. Musa Qala's reputation exceeded Sangin's in terms of gnarly nastiness but nothing could have prepared us for the Taliban welcome party we were about to encounter.

Chapter 18

We waited for the AH to position on the HALS.

'Tower, this is Wildman, request reposition HALS.'

'Wildman, Tower, reposition HALS, clear take-off at own discretion with Doorman 26.'

'I can see the AH moving,' said Dan.

'Roger, I can see them lining up,' I replied.

'Pre-take-offs are good, ready to lift.'

We were now lifting into the Afghan pitch black. My eyes were adjusting to the reduced peripheral vision. I concentrated hard. Lifting was tricky as I was not used to the reduced visibility. As the aircraft climbed up into the inky blackness, my eyes relaxed, the green hue of the goggles became more familiar and as we advanced into the cavernous expanse of the night sky I settled into the flying. There was an eerie silence in the cab. The normal banter was missing. We were all pissed off about losing the T1 and here we were again, not half an hour later, heading out to pick up another casualty.

We routed north and ahead I could see the outline of the glacier mountains of Bagram, silhouetted in green by the light of the NVGs. We had a temporary respite at the safety of height but it was to be short-lived because we all knew that we were soon to be dropping down to low level and facing an unknown quantity of Taliban objection.

'Where's our little friends,' I asked, referring to the Apache.

'Low at the five and seven,' came the reply from Sam.

We were coming up the eastern side into Musa Qala DC,

which was smack bang in the middle of the town. As we let down to low level, I said: 'Right, gents, same as before: I am Mark, this is Daffers, you are Sam, you are Dan. Are we clear?'

We all knew that we were back in the zone once again. 'Descending,' I said.

The ground commander wanted us to come into Musa Qala from east to west, along the length of the town.

'Doorman 26, this is Wildman, ICE over,' said Dave on the radio.

As we descended it wasn't too bad. We had dropped down to low level by the outskirts of the town. But as we came over the edge of the town all hell broke loose. The tracer and small-arms fire started coming up towards the cab. It seemed that only a few seconds later everyone in the world wanted to kill us and we hadn't even ID'd the landing site. On the one hand we could see that it was kicking off around us but at the same time we needed to ignore it to focus on bringing the aircraft into the LS, without flying into the ground.

Daffer was talking me towards the target.

'Two Six Five, six miles,' said Daffers talking me to the LS. 'We have got five miles to run . . . Come left, come left, come left, come left. Roll out there. There it is twelve o'clock, twelve o'clock . . . It's on the nose. Twelve o'clock. Can you see the light?'

'Visual. I have got the signal,' I replied. 'Pre-landing checks.'

'Holds out, CAP clear, Ts and Ps are good. Brakes off, swivel switch locked. Bugged at forty feet my side,' said Daffers, short, sharp and to the point with no unnecessary blah.

Dan was on the doors and Sam was on the minimally armoured ramp firing the M60. The defence equivalent of sitting on a big platform made out of tinfoil. 'Anyone got a bacon sandwich and a big fuck-off torch so that I can really make myself feel well defended and conspicuous,' muttered Sam, in one of his inappropriate joke moments.

Dan was on the right on the number two gun.

'I am just going to have a look on the left-hand side,' said Dan. He turned around and as he looked out on the left he saw an RPG being fired from a building beneath us. It looked like a firework and was moving at normal speed, but to him it seemed like it was going in slow motion. He looked at it, thought it was going to be close but not actually going to hit us so he didn't call a manoeuvre. He simply opened fire and began hosing down the building with bullets.

Sam saw where Dan was shooting, took the target over and began putting fire down on the building. While that was happening Dan looked out again and saw another RPG being fired. On the radio Dan called: 'RPG' in a very calm manner and pointed his smoking M60 in that general direction. Then Sam down the back called: 'RPG' in a more determined way and he continued firing. The RPG flew past the ramp and missed us by a mere 3ft. Sam felt the heat of the missile on his face as it whizzed past.

'Fucking hell that was close,' I heard a sardonic Sam mutter.

It was amazing how calm we were all being in the face of RPGs that were flying past the aircraft. RPGs are designed to ruin our day. They are 5ft-long, highly explosive anti-tank warheads launched from the shoulder. They were the weapons that blew the tail rotors off the Black Hawks in Somalia and are a nasty, easy-to-use piece of kit, posing a real threat to the aircraft. I had experienced crewmen more stressed in bad weather then Dan and Sam were with certain death missing us by a cat's whisker.

Sam turned the gun and started hosing down a house. Almost instantaneously someone else opened fire on the left.

Dan immediately returned fire. He was worried that if it was going off on the left what would be happening on the right.

'Shit!' shouted Dan. 'Checking right now!'

He opened fire. I heard Dan say over the intercom: 'Target neutralized.'

He turned round and moved to the right side of the cab and watched Armageddon unfold all around us. The scene was one of carnage as bullets, RPGs, mortars and ordnance were exploding all over the DC. We were under intense and heavy fire. Dan and Sam were firing constantly at the enemy positions that were firing up at us. Dan switched between the right and left gun. The lads had clipped two cans together – they went through 500 rounds each – and were constantly reloading the M60. We were literally fighting our way in. We had just flown into Judgement Day. It was full on.

But then all of a sudden there was a shudder through the airframe. Daffers and me looked at each other. Daffers said: 'I think we have just hit a bird.'

'Oh yeah, I think we have,' I replied.

We thought nothing more about it.

The tracer wasn't very clear. But occasionally I could see flashes. Daffers was really switched on. It was a black as pitch. The points of reference were mud hut, mud hut, mud hut, slightly larger mud hut and the HLS was being indicated by a lone soldier sending up a light signal which could only be seen through goggles. We were the most at risk that we would ever hope to be in a Chinook.

We were 80ft above the ground. We were slowing down. The alarm bells were ringing in all our heads. The deceleration increased our risk as a target, and if we were going to get shot down it was going to happen now. The only thing so far that had prevented the Taliban from hitting us was our accelerated descent into the LS.

As we slowed down, I looked at the LS and I thought: *I can get in. I can do it.* I had ID'd the LS. At the same time I was throwing the aircraft around so that the Taliban were unable to get a fixed aim at us, all the while trying to get my landing timed perfectly. It required every ounce of my concentration. I brought the nose back and focused on getting the right attitude.

There was a flash of clarity in my mind. I felt I had nailed it and could land on. The Apache was talking to the ground call signs. The ground call signs and the AH all at once started shouting over the radio:

'ABORT! ABORT! ABORT! ABORT!'

Shit, I had it fucking nailed. Fuck, fuck, fuck! I said to myself.

I rocked the nose forward and pulled power. I accelerated away from the LS. I flexed my left arm and started pulling up on the collective. It took a couple of seconds for the airspeed to build and I heard the engines bellowing as the power increased.

As the nose dropped I could see the horizon rising on my goggles. Daffers was constantly monitoring the engines to make sure that we didn't overheat them and blow them up.

'We are overshooting.'

'No shit.'

As I said this I pulled in more power and steered the cab into the ascent. We flew over the top of the LS. It was the best visual we had had on it to date. 'Oh, there it is!' said Daffers, as we pulled away from it. I turned left, manoeuvring the cab like a bucking horse. I was changing my position so that the enemy couldn't get us in their sights.

I really hoped that we would get out of there without being shot down but I knew it was going to be a close-run thing.

'Engines are good,' said Daffers.

I pulled up more power to keep the speed on. We climbed

at an accelerated rate of climb to several thousand feet. Daffers scanned the horizon for the Apache. The last thing we wanted after getting out of that shitfight by the skin of our teeth was a mid-air collision as we were ascending.

I called the Apache over the airwaves: 'Wildman, Doorman 26, request loc stat' – their location status.

'We are holding to the north.'

As he responded on the radio I could barely hear him against the rattle of his guns as he continued to mallet the DC.

There was tracer coming at us from all directions. But we thought we had taken rounds through the ramp at the back. The Taliban started to mortar the site. If we had landed on we would have been besieged by small-arms fire, machine-gun fire and a mortar bombardment. There was no way the Chinook even with its extensive defence suite and armour could have sustained that level of bombardment without exploding.

We pulled over to one side and as we climbed away watched the firefight on the ground go off. It was not until we reached height that we could relax a little, once we knew we were clear of any rounds or bang and we knew that the skies were clear of aircraft.

On the ground they could see all the rounds coming up at us. In the sky the Apaches could see all the rounds coming at us. It was more important that we didn't lose a Chinook than it was to pick up the casualties. If the Chinook had been shot down it would have become a death magnet.

Sam and Dan were checking the cab.

'I saw tracer go underneath. I am pretty sure that had hit the ramp,' said Sam.

'I think we took rounds,' said Dan.

Everyone in the aircraft was very subdued. This was a shit day. Not two hours earlier an IRT that we had picked up had died and now we had just failed to get into to Musa Qala. I was not a happy bunny.

'Is everyone all right?' I said.

All the instrument panels on the front and the maintenance panel down the back, which told us how the engines, hydraulic systems and transmissions were operating, were indicating that the aircraft was ok.

'Do you know what?' I said. 'We ain't going to get in there right now.'

'Yep, that's not going to happen,' replied Daffers.

Despite being pissed off that we hadn't made the landing, with all hell breaking loose beneath us we had no doubt that it had been the right call. It was the ultimate no-shit statement but it had to be said. I called Dave on the radio. 'We are going to have to take this home and rethink it,' I said.

'Roger that,' he said.

'Let's talk about it back at base,' I said.

'Roger, mate. That was a bit fucking sparky!' he said.

'You can say that again,' I replied.

Dave, in the Apache, stayed on until he was almost out of fuel to give the guys in the DC as much back-up as possible. We had obviously driven a stake into the middle of a wasps' nest and it was all kicking off big time at the DC.

We sped back at about 120 knots. It was important that we gave our reapproach some careful consideration. Dan was on the number two looking at the ramp and in the distance he could see the flashes lighting up the sky.

'It's like the Disneyland fireworks back there,' he said.

Sam pitched in. 'Fucking hell that was as lively as you like!'

'We are going to have to rethink this. One of the T1s has got a gunshot wound to the neck so it's likely we are going to have to go back in,' I said to Daffers.

'Yeah, mate,' he said.

We were flying fast to get back so that we could work out how we were going to get in there again. It pissed me off that we hadn't done it first time.

'I thought I had it nailed as we were coming down,' I said to him.

'Don't worry, mate,' he reassured me. 'It was going off like a frog in a sock.'

As we were flying back into Bastion we could see in the middle of the desert the conglomeration of lights twinkling in the distance. The night was pitch black but the base was lit up and could be seen from twenty miles away. It was getting quickly bigger and brighter as we drew quickly towards it.

'Bastion ahead,' said Daffers.

'Yep, roger, got it,' I replied.

On arrival at Bastion we spoke to Tower on the radio to get clearance to come back in.

'Doorman 26, do you require Nightingale?' Tower replied.

'Tower, Doorman 26, negative, normal spots.'

'Roger, cleared to land spots.'

There was always a sense of relief when we crossed the fence at Bastion, almost like we had come home and were safe now, especially once we had shut down. The approach into Bastion can sometimes lead to complacency. A couple of people I know have just missed other aircraft departing the base, as they mentally switched off approaching the fence. The truth is it ain't over until the engines are shut down and the aircraft is signed back in to the engineers.

We made our approach to the landing spot. Unfortunately Bastion was designed in such a way that in order to approach into wind, we had to park facing the other direction and couldn't fly over the camp to land. It was a ball-ache and it meant we had to come to a hover over the landing spot, then kick the tail round through 180 degrees, blowing the shit out of everyone and everything within a couple of hundred metres of us. As soon as the wheels hit the ground and compressed we, and the aircraft, started to switch off.

'Stabs' – the AFCS, or autopilot – 'are out, brakes are on, EAPS off, clear PU.'

I reached up and flicked the switch to start the APU, which keeps the aircraft systems up and running when the engines are switched off. I heard the whine of the APU in the background.

'Good PU,' came the call from the ramp area.

'Roger – shutting down,' I said.

I pulled the ECLs – throttles – back out of the flight gate above my head and the rotor blades immediately began to slow, without the engines to power them around. Daffers informed me that he had switched off all the self-defence kit and navigation kit.

'Rotor brake is coming on,' I said, as I pulled down on the massive rotor brake handle – it needs to be big to stop the huge blades from spinning round.

Once the blades had stopped spinning we told Air Traffic that we were shut down and complete. Daffers added a 'good night' on the end. I am usually less polite when I am ballbagged.

With that I switched off the APU and the silence was beautiful. Flying helicopters is a noisy business, living in a war zone is a noisy business – there was very little peace in our lives out there. The moment when we shut down the cab completely was one of those rare moments I liked to savour. But not for long as there was still business to do. I was keen to get to the CP to work out our next move.

'Right, Daffers, you and me, we'll go and speak with Mike and the Colonel at the CP. Dan and Sam, you stay here with the engineers and give the aircraft a quick once over,' I said over the intercom.

I reached up and flicked off the battery. I took my helmet off and rested it on the centre consol. My head was roasting hot from being in my helmet for hours. I wiped my head with my hands, and it cooled quickly as it was freed from its casing. This was one of the few advantages of being as bald as a baby's backside.

I climbed out of the cockpit, into the gangway, and walked through the aircraft and down the rear ramp. The night air was cool and I reached into my pocket, pulled out my Marlboro Reds and lit one. Dan, who didn't smoke, and Sam joined us. As a crew we all had a Red and took a moment. It didn't last long and nobody spoke. Sam and Dan went back inside the cab to get on with the tidy-up. I was stiff from being in the seat for so long so I stretched out my arms and rolled my head around my neck.

While Daffers and I were waiting for the Landie to turn up to get us back to the CP, one of the doctors from the MERT came over. He was an Army Lieutenant Colonel. His face was ashen because it had been a bit of a hairy night. I lit up another fag.

I took a long draw on it, savouring the numbing of the head rush as the smoke infiltrated my bloodstream. My mouth was dry and claggy but the bittersweet flavour of the saltpetre and tobacco was what I needed. My heart started to pump a bit faster and the fag, combined with the adrenaline, gave me a little lift.

'Have you got a smoke, mate?' asked the doc.

'Sure thing.' I reached into my pocket and pulled out the Reds. 'Here you go.' I handed them and my lighter over to him. I looked at him and said: 'You do realize that we are going back.'

He looked back at me. 'Yeah. I thought as much,' he said. With that he took the longest drag of the fag his lungs could manage, threw it on the ground, stubbed it out and walked back on to the cab.

The Landie arrived and we drove as fast as we could to the CP.

Dan and Sam stayed with aircraft. The cab was in a shit state inside. It was full of bullets and cases and was still covered in blood from the Sangin IRT.

The engineers opened up the casing that surrounded the centre hook and inside it was masses of congealed blood. They had about twelve kits suitable for cleaning blood off a hand wound and needless to say they went through the kits really quickly but barely made an impact on the blood. It was pretty obvious that we were really lucky not to have lost that cab. A young engineer made a really useful comment to Sam: 'If you'd been flying that a bit longer you would have all crashed.' It went down like a shit sandwich. Sam was ready to punch the young eng but Wilf, the JENGO – Junior Engineering Officer – came over and said to Dan: 'Mate, that aircraft is not going anywhere.'

Dan looked at him, slightly drawn in the face. 'Can you start to prep us another one?'

'No problem,' he said and went off to sort out another cab.

Chapter 19

As we walked into the 3 Para CP tent, Tootal, the RSM and the Ops team, McGinty and OC 'B' were stood around the bird table. We were still buzzing. I was pumped up from the experience but I knew that it wasn't over. I knew that we were getting back in a cab and going back out there to take it on again. It was all I could think about. In my head I was playing it out again and again. How could we do it better? How could we get in there quicker?

One of the 3 Para officers, Captain Swann, who was the Air Liaison Officer, the link between JHF(A) and 3 Para, came over. 'How was it, mate?'

'Interesting to say the least,' I replied. 'We've gotta go back. We didn't get them.'

Swanny was an on-the-ball chap. He was totally clued into the requirements of getting troops on and off the cab quickly. Everyone at the JHF(A) rated him.

On Sam's first Musa Qala IRT pick-up, a doctor had wandered on to the cab and been brought back to Bastion by mistake. He had come up with the SOP that guys who were to get off the cab wore snapped, fluorescent Cyalumes so that the crewmen could clearly see who was who. The rule was that anyone without an IR Cyalume stayed on, anyone with one got off. It was an easy procedure to administer and it demonstrated Swanny's understanding of and empathy with our needs.

He asked us: 'Gents, what do you want the lads on the ground to do as you come in?'

Colonel Tootal walked up to us and interrupted. 'Right, what's happened?' he said.

'We routed to the north and then dropped in from the east as advised,' I said to the room.

A J2 bloke from the 3 Para Int team piped up. 'What were you doing going in from the north-east?'

'We were told to go in from the north-east by the ground commander.'

'That's where all the Taliban heads of sheds live. They have got their own security. You must have come past all of their strongholds. No wonder they all had a go at you.'

The look on Daffers' face said it all and I agreed with him: *Why the fuck didn't we know that before we went in?*

Colonel Tootal turned to me and said: 'So what do you want to do?'

'Well, not coming in from the north-east again would be a good start.'

Everyone sniggered, even in the height of the tension we were able to crack a smile. The door opened and Sam and Dan walked in and stood quietly at the other end of the bird table.

'I want my lads but I am not willing to lose an aircraft so I am not going to ask you to go,' said Tootal.

I looked around at each of them.

'What do you think?'

Daffers, Sam and Dan all answered simultaneously: 'Yeah, let's do it.'

It was a horrible Hollywood moment. It wasn't meant to be deliberately cheesy.

Tootal asked one of his watchkeepers: 'What's the state of the casualty?'

'They are trying to stabilize him and buy us more time.'

'I need some medical advice. Have we got four hours?'

'I'll get back to you, sir,' said the watchkeeper.

'What are your thoughts?'

'We'll throw everything we need to at this. I think we need to come in from the north-west, sir.'

Unbeknownst to me, contingency plans had already been discussed between Tootal, McGinty and OC 'B' while we were routing back to Bastion. Tootal talked to the Royal Artillery officer.

'What fires have we got?'

'We have asked for a Spectre, sir.'

The CP had made contact with our American friends to see if one of their AC-130 Spectre gunships was available. They have been around since Vietnam in many different forms. In its current form it packs a tremendous amount of firepower – a Hercules armed with a 105mm howitzer, 20mm cannon, chain guns and Miniguns – all radar-laid and all-weather capable so there is literally nowhere to hide. The saying is, you can run but you will just die tired. They really are the ground troops' favourite as they can stay up for hours and have a lot of ammunition on board. The firepower that they can deliver on to a target is truly awe-inspiring and I would never want to be on the receiving end. They can destroy buildings, tanks, vehicles and people with ruthless efficiency. It was a very impressive and capable aircraft. The Americans said: 'Maybe.'

Hoofing! I thought. *Bring it on.*

'And we have the MOG in support.'

This was 105 Light Artillery positioned out in the desert.

'We are trying to find out what CAS' – close air support – 'is available.'

Tootal turned to me. 'We need to cover the sound of our approach.'

'Looks like we need pre-emptive fires,' said OC 'B'.

Normally we would enter the LZ, wait to be shot at and then open fire. That was the norm. Any preliminary fire had to be cleared at higher level. All firing done in Afghanistan is legal and has been cleared by lawyers. They were operating under ROE 429 A – which meant that we could shoot at them if we could identify them as the enemy (allowing them to engage any identified enemy). Some places in Afghanistan were in 429 B, which meant that we could only shoot at the enemy if the enemy was shooting at us. Once we were in a hot contact all bets were off.

But if we wanted to shoot at people pre-emptively then the targets had to be cleared. We were required to get cleared target sets: how far from civilians? How many civilians could be affected? What was the collateral damage level set? How many civilians are we prepared to have killed in collateral damage for this assault?

The three military principles on this were discrimination, military necessity and proportionality. This meant we had to use a proportionate amount of force, and use it discriminately – if a stiletto will do, use a stiletto and not a 2000 pounder. This meant that we had to use the minimum amount of force necessary to achieve the military necessity. The military necessity was often: is someone shooting at you? The situation became more complex if they were not shooting at you. How was the military necessity defined under those circumstances? What is the military necessity for dropping this bomb? Is this size of bomb proportionate or do we use a smaller bomb? Are we using the smallest bomb necessary? Will it be effective and achieve the military necessity? Is it proportionate? Do we need to use a bomb? Could we do it with a sniper's rifle?

McGinty had a chat with Colonel Felton, the CO JHF(A) back at KAF.

'What do you think?' Felton had asked him.

'Sir, I think this is the most dangerous mission that British rotary aircrew have ever been asked to be engaged in,' said McGinty.

We were flying in to pick up casualties from a known landing site where the enemy could predict our arrival and we could do very little about it other than choose not to go.

'And, sir,' he added, 'if we carry on like this we will lose a Chinook. The clock is on. It's not a matter of if, it's when.'

It was the big known Taliban objective to take down a cab. Politically, it would send a very strong message to the extremist world to have a downed Chinook blazing across the Al Jazeera network. A helicopter is a flying machine. They are an omnipresent sign of power. Shooting one down would represent a significant morale victory for the insurgents. Good juju for the bad boys.

On the phone to Felton, McGinty said: 'Sir, I don't think we should do this. It is ludicrous. They got shot at too much to go in and land. They had an RPG above and RPG below. This is not good. Look, sir, I don't think that we should go in there and do this again. I think that we should take a different approach. CO 3 Para is now talking to Brevat Red and I think we need preliminary fires to do this.'

To us not going back in was not an option. At the bird table, McGinty was asking Tootal the difficult questions that needed to be asked.

'Are we prepared to lose a Chinook for this casualty?'

'No,' said Tootal.

'I agree with OC "B". I think we need preliminary fires. If we have that we can get in,' I said.

As we were all in agreement on the strategy, Tootal sought authority higher up the food chain to conduct preliminary fires.

The plan was taking shape. We were going to route in from

the north-west, leaving clear a small avenue of airspace for the Chinook to get into the DC. We were going to shoot our way in. Instead of waiting to be shot at we were going to put down fire. In the middle of all this McGinty was looking at how many Apache aircraft and crews were available. After the events of the day – at Garmisir, Kajaki, Sangin and Musa Qala – the Apache crews were on the bones of their arses. Every Apache crew in theatre had flown over maximum hours apart from McGinty and his back-seater. He was looking at who had flown the least. All he could do was fly everyone who had flown the least. We were scheduling four Apaches in the escort this time.

This meant that McGinty had to step up to the plate and fly the last sortie back into Musa Qala. Normally what would happen with a pair of Apaches was that you would have one watching the Chinook and one watching the LS. We might vary the exact timings. Sometimes we would try to arrive ahead of them, but we realized that in order to arrive ahead you need to get there really early and start malleting things or arrive thirty seconds ahead, which could give the Apaches just enough time to get the situational awareness and say: 'Yes, this is good. Land on.'

One of the pair of Apaches was to trail us and the other pair were to be used to run in and take targets from the ground controller, the widow. We wanted to use the sound of them firing off their 30mm cannon to cover the sound of us coming in.

The priority was to get us on the ground. Everything was being put in place to ensure our safe passage. We still hadn't had confirmation on the Spectre but the fires officer had confirmed that an A-10 was definitely available. The A-10 Thunderbolt II, nicknamed the Warthog, is an American single-seat, twin-engine, straight-wing jet aircraft flown by the US Air Force exclusively to provide close air support for ground forces by attacking tanks, armoured vehicles and other ground targets, with a limited air interdiction capability. It is equipped with a 30mm GAU-8 Avenger Gatling gun, capable of firing around sixty-five rounds per second, a highly effective tank-killing weapon. The myth has it that firing the Avenger cannon actually makes the aircraft slow down in flight. Ground troops love it as it can fly slow and low and has massive firepower available. The pilot sits in a titanium bathtub and the aircraft is designed to be able to soak up small-arms fire – in fact it can still fly with one engine, one elevator, one tail and half a wing ripped off, so it is the perfect weapon for close-in killing.

Although the situation was major, the atmosphere wasn't tense. The humour and the banter still flew. It was more like: *'Fucking hell we are gunna go and do this.'*

To some extent it was just another day at the office. Let's roll our sleeves up and get on with it: *'We are not here to fuck spiders after all.'*

I never thought we were going to die. Pilots never think that they are going to die partly because it is a well-known fact that once you are sat in the cockpit of a military aircraft you are invincible. As absurd as it sounds, there is an element of truth in that.

The reason that aircrew are up their own arses is because flying is very empowering. It makes you master of what you slay. You make this thing fly. It is ego fuel for a good reason, because you are sitting there and this is where you strap into your seat. This is the seat where I make big things happen. This is my office. Maybe you don't really feel invincible but what you feel is it's not going to happen to me, so it's ok. We just need to go off, get this bloke, get it done, get it done properly, come back. Confidence was high. It had to be.

The run back into Musa Qala was all about mitigating risk. The A-10 was definitely going to be flying there, the Apaches were going to be firing and 3 Para on the ground were going to be firing.

With the plan formed, we now left 3 Para CP to go through a formal aviation brief in JHF(A). Before that we had a quick fag break. Sam, Dan, Daffers and I convened outside the JHF(A) tent. 'I know the answer, but I have to ask. Is everyone happy to go back in?'

'Slim chance of success, certainty of death. What are we waiting for?' said Sam.

'All right, Gimli, get your war axe out.' We all chuckled. 'Seriously, is everyone happy?' I asked.

They all nodded. I felt better. As captain of the aircraft, I wanted to be sure that outside the spotlight of the CP they were genuinely prepared to go back.

We all walked back into the CP. Mr Mack, the chief crewman, a wily old dog, salt of the earth, came up to me. He had been on the squadron for years, he'd been everywhere and done everything. 'I want to take Sam off. I need him up at KAF,' he said to me.

'No way, mate. He needs to stay on. I don't want him off the crew. I want to keep him on. He's flown in with us. He knows the lie of the land,' I said firmly.

I knew Sam wouldn't want to be taken off the gig. We started this as a crew and we needed to finish it as a crew, no matter what. If I knew one thing for sure it was that.

'Well, I was going to put Spence on with you,' he replied.

'Let's put Spence on as well anyway. It will be better to have him onboard. It was going off like the dance of the flaming arseholes on the way in,' I said.

'Ok, roger. You can keep Sam but take Spence as well,' he agreed.

This meant that we would have all the guns manned at all times. It was for the best.

The AH crews, OC 'B', Spence, Daffers, Dan, Sam and me gathered round the bird table. I led the brief, which served to confirm all the details of the plan we had established next door.

I covered the order of lift. Two Apaches were to go up first, then myself with the other two tailing. It was comforting to know that I would have two Apaches following me in.

The most important call for us was whether we were cleared in. We had been using Cherry/Ice for Hot/Cold – hostile or safe – but I decided to change it.

'Cherry/Ice call will be Mowlam/MacAndrew.'

'What?' There was a look of confusion on the faces of AH.

'C'mon, lads, think about it. Mowlam you wouldn't, MacAndrew you would.' We all laughed.

One of the guys from Tootal's tent came over.

'Sir, the casualty has stabilized. He can hold until midnight if necessary,' he said.

I looked at my watch. 'The time is now 2130. Let's leave at 2300,' I said.

Chapter 20

I was tying up loose ends before heading back to the tent. One of the watchkeepers walked over to me and interrupted our conversation.

'Excuse me, sir, the Line is on the phone. They are asking for you,' he said.

I walked over to the phone and picked up the handset. 'Major Hammond speaking,' I said quite abruptly.

'Hello, sir, it's Wilf. The aircraft is U/S. You have taken some rounds through the front head. It might be worth your coming and having a look at some point. We are prepping the spare. You will need to come down to the cab to transfer your kit over,' he went on. 'There were seven bullet holes in total in the aircraft. Six on the actual airframe and then one almost through the vertical hinge pin.'

The round had gone through less than two inches away from the vertical hinge pin. Two inches closer and it would have taken the pin out and disconnected the rotor. The entire 30ft blade would have been thrown off the aircraft, which would have pitched the cab into an uncontrollable spin, and ultimately resulted in all our deaths. It was that goddamn close! There are very few things that will instantly down a CH-47 as it is an incredibly resilient cab. Losing a whole blade is, however, one of them.

I put the phone down and looked across at Daffers.

'Mate, we didn't hit a bird. We took a hit! The cab is U/S. It's totally riddled. A round nearly sheared through the vertical hinge pin.'

'Shit!' said Daffers, quite shocked. 'There but for the grace of God . . .'

But there was no time to waste thinking about how lucky we were not to have crashed and died.

The aim was to come in this time from the north-west across the wadi and over the hill, avoiding the Taliban leaders' houses and their increased security. The pick-up was for four casualties. The rules had changed: what we were going to do to the village was rip it a new arsehole. The fire was going to be pre-emptive. The plan was to mallet the place before we went in so that we could land the cab as safely as possible.

I felt good about the plan. There was so much firepower going in that I knew they were going to get malleted. My biggest concern was that I didn't want to pork the landing and lose the cab. The implications of losing a cab were enormous. It would be like flies around shit in terms of the attrition because we would have to send another cab to pick up all the crap caused by our cab going down. It didn't really bear thinking about.

I felt like I had drunk five cans of Red Bull. I chain-smoked. It was as if I had just come from the gym, the endorphins racing around my body like atoms. I had no doubt that we would pull this off. I never thought we were going to die. It just didn't cross my mind.

I was focused on not repeating the errors of the last approach. It was important that we came in from the right direction. I had confidence in the assets that were in place and I knew they were there to protect us. I felt quite bullish. I was less concerned with the Taliban's losses. If they were going to attack Brits, then they were going to get malleted. They attacked us and we had to respond. We had to pick up the casualties and it was the Taliban's fault that we had to collect those casualties. We knew they were desperate to bring a Chinook down and we weren't going to make it easy for them.

My whole philosophy was that we were there to protect people rather than kill people. If they were corrupt like the genocidal warlords, the drug barons and the Taliban, and they killed Brits and then died as a consequence of their actions, then that was their fault. If they wanted to fight then we would take the fight to them.

Getting the cab in and out without incident was my priority. The most important thing was the landing. On the previous two engagements I had been lucky and I had been able to get in and out without losing the aircraft. For me all the pressure surrounded the landing and making sure that I didn't pork it up. If I did I could end up killing everybody and that outcome was completely unacceptable. Everything was at stake, people's lives and my reputation as a pilot.

Chapter 21

We headed back to the IRT tent and tried to get some sleep but it was impossible. The guys wouldn't settle down. I said to them: 'You are meant to be asleep.'

'All right, Dad. It's not as easy as that!' replied Sam.

'Fuck this, I am going for a shower,' said Daffers. About ten minutes later he returned, wrapped up in a grotty Coventry FC towel.

'Nice towel, mate,' said Sam, smirking.

'Mate, do you want to borrow some of my menthol powder for your balls? Freshen you up a treat,' I asked him.

They took the piss out of the menthol talcum powder I used on my balls. It was, however, the best solution for the prevention of scrot rot in the heat of the Afghan desert.

'No thanks, mate. I am ok,' said Daffers.

'Suit yourself, mate, suit yourself,' I said. 'But don't come crying to me when your nuts are rotting away, corroded from your own body's sweat.'

Dave's crew were in the tent with us. They were asleep and blissfully unaware of what we had just been through. We all tried to get our heads down to get our essential crew rest but none of us could settle. We were surging with adrenaline. We had unfinished business to do.

On the one hand it was weird having them sleeping around us. I had the urge to shake them like an over-excited child and tell them how it had just fucking gone off at Musa Qala and we were going back in, but then again at the same time I didn't want make a fuss. I wanted to maintain my cool.

Dan and Sam both thought we were going to end up in a smoking hole in the ground.

'Mate, I think we are curtains,' said Dan.

'Don't be stupid,' I said. 'No we're not.'

'I don't care if we are,' said Dan, 'because I am telling you that I would rather die than look that kid's mum in the eye and say: "I am sorry, missus, but I couldn't go back and get your son because I thought I was going to die." That to me is worse than ending up in a smoking hole in the ground.'

Both Dan and Sam seemed to have come to terms with the idea that they might die, but if they did they would at least die trying. By mentally preparing themselves for death they were empowered because strangely they were no longer afraid of it.

I looked up and Dan and Sam were sat on their racks writing goodbye letters to their families. I silently watched them as they placed them under their pillows. I never mentioned it to them and they didn't talk about it to me. They had to do what they felt was right. For each of us the preparation for our return to Musa Qala was a very personal journey. I still didn't think that we were going to die but I knew it wasn't an easy task that we were about to undertake.

Sam had gone outside to have a fag, spinning a dit with an Army Air Corps officer. I wandered over to join him and pulled out the Reds.

'Fucking hell. We are going back into Musa Qala! Part of me can't quite believe that I am doing it,' he said.

'It could be worse, mate,' said the Pongo officer.

'Could it?' said Sam. We both looked at him a bit perplexed.

'Yeah, mate – you could be the Taliban,' he said.

'What, you mean have bad facial hair and be clothed in a sheep,' said Sam.

We all laughed.

It was a fair point. The re-entry in Musa Qala was going to be as magnificent as the *William Tell Overture*. Spectacular, to say the least. If it was going to be our day to die, we were going out with a bang. If we were going to fail, we were going to fail magnificently.

We jumped back in the Land Rover and headed back to the cab. When we arrived the MERT were already in place. They had cross-decked their kit. The engineers were there giving the cab the final once-over.

I walked up to the back of the aircraft. The cabin lights were on and the MERT were silently getting their shit in one pile, hanging up drips. It felt a bit tense. I could sense their apprehension. It was harder for the guys in the back as their destiny was in my hands. They knew they were going to fly into a shitfight and they had to cross their fingers and hope that they would come out of it alive. It was enormously brave of them and those guys had all of our respect. I walked past them up to the front and started to check my kit, which Sam had cross-decked for me on to this aircraft.

We had three HRF cabs up there and we were now on our second. Daffers did a walk-round. We jumped on, and started up the engines. A rapid start. We went through the start-up procedures.

In peacetime, we take around thirty to forty minutes to start the aircraft and get ready to launch. This is because of the many complex systems involved. In war zones we normally do it in around fifteen minutes. When I am in a hurry and people's lives are at stake, we can do it in under ten.

As soon as the APU was on, I started to fire up the engines. Daffers had jumped back in and was working like a one-armed paper hanger with the nav kit and self-defence suite. With the rotor brake off the blades began to accelerate and were soon thumping through the warm night air. As I pushed the throttles forward the aircraft began to strain at the leash – it was as keen to get back in the fight as we were.

I gave the MERT a quick update of what was going to happen.

'Is the doc on comms?'

'Yeah, I am here.'

'We are going back to Musa Qala to get the T1 and the T2s.'

'Roger, that's all understood.'

'I haven't got any more information at the moment. We have to go. You need to know the ride in could be interesting.'

It was my way of telling them that we were expecting it to be a little bit sparky.

We then ran through a quick set of take-off checks to make sure nothing had been missed in the rapid start, and that we had everyone onboard that we needed.

The first set of Apaches rolled along the HALS first and then we took off behind them and were followed by two more. The first two were going a slightly different route to us and they were headed north-east straight to the target to begin the malleting. We were heading straight north so we were deconflicted. We had already briefed that we were going to be at height and they were going to be slightly lower.

As we were flying in towards the site from about twenty-five miles away we could see that that Musa Qala was lit up! I thought it was the AC-130 gunship hammering the crap out of the place. We didn't know at the time but the AC-130 wasn't available.

Unbeknownst to us at the time, the A-10 had come into the western side, where the Taliban had stored all their weapons in an ISO container. The A-10 had ripped it up and it was blazing so brightly that it illuminated the town.

'Mate, can you see that over there?' I asked Daffers.

146

'Yeah!' he said.

'I think that's the AC-130 indicating the target for us . . . Nah it's too quiet,' I said.

On the way in we were on goggles. It was completely surreal because we could see missiles coming off the rails of the aircraft, hitting the ground and exploding.

'Blimey, they really are giving these guys what for,' said Daffers.

A huge light was illuminating the sky.

I said: ' Fucking hell, mate, we are going to have to get them to turn that off because it'll fuck up the gogs,' still thinking it was the AC-130.

'Mark, look under your goggles, mate,' said Daffers.

I looked under my goggles.

'Shit, that's fire!'

Daffers said: 'Fucking hell – they are getting a malleting.'

'Fucking hell, that's going to close our goggles down it's so fucking bright,' I said.

As we flew towards it, it was like flying in daylight. It lit up everything. I was detached from the whole experience at height in the safe little bubble, watching all hell unfold beneath us. Although we were tense, at the same time it was so surreal that the atmosphere in the cab was almost light-hearted. At that moment in time we felt protected from the Armageddon ahead of us. In a strange way we were removed observers and in our happy place all looking out of the windows thinking: *Oh look, it's really kicking off down there.*

Dan was on the intercom.

'Oi, Mark.'

'Yeah, mate,' I said.

'Natasha Kaplinsky, Nell MacAndrew, Davina MacCall,' he said.

Even at this time my crew were still joking. I thought

quickly but it wasn't an easy one because quite frankly I would do them all. To some it might have seemed an inappropriate moment to play Shag, Marry, Kill, but that is the aircrew way. In the face of adversity to take your mind off the enormousness of the task ahead you play something really trivial and inane. If you are transiting in the safe zone at height thinking about whether you would shag or marry Natasha Kaplinsky you are not thinking about flying into Armageddon. It keeps it light. It keeps you focused on the job and not the threat.

'It's a tricky one, mate. Err . . . Marry Natasha, Shag Nell, Kill Davina.'

'Really! Poor Davina, she's a honey. She'd get it. Nell would have to go,' said Dan.

It wasn't until we began our descent that I realized: *Shit, I have just killed Davina MacCall!* It was too late to change my mind.

Musa Qala was lit up like Chinese New Year. On the goggles we could differentiate quite clearly the cannon going in, Hellfire missiles being released, and the rockets going off.

All I could think was: *Fucking hell, THAT is where we are going!*

It was nearly midnight, the darkest time, so the contrast was stark. It created the perfect backdrop to illuminate the incendiary arsenal of firepower that was being unleashed in preparation for our arrival. Our recollection of the experience would be *for ever* indelibly imprinted on all our minds.

A little bit further out than our previous descents, around the time we went over a wadi, I gave the unofficial signal to assume positions over the intercom: 'Ok, gents, from now on I am Mark, this is Daffers, Sam, Dan and Spence . . . Descending,' I said.

The crew assumed their positions. On the way in initially it was fairly quiet. All around the Apaches were pounding the shit out of the place.

We had already test-fired the guns on the transit to make sure that they were still working. It was important for us all, including our passengers, to know that the aircraft guns were functioning well. They might well be called on to save our lives and we made sure they worked for that exact reason. If there was a problem we could always fix it at height, before we dropped down into the maelstrom of low level.

Once we had begun our descent the crew knew implicitly that the job was on. We internally braced ourselves and mentally stepped up to the plate. I didn't have to say anything. Training and experience had taught us what to expect from this point on and we just silently moved into game-on mode.

It did seem as though we were flying in slow motion. The entry into Musa Qala was like the opening scenes of *Terminator II*, with fires burning all over the place and armed soldiers, heads down, scurrying between them in the wreckage. We dropped down in slow motion into carnage and destruction. For a split second I thought: *Are we really here? There is just stuff burning everywhere.* It felt like we were gliding in at 10 knots, watching the mayhem below.

We were coming down to low level, over the high ground, and looking for the target. We came down reasonably quickly. As we were doing so we were constantly eyeballing the target. The ISO container was burning so brightly that it was bringing down the effectiveness of the goggles so it was very difficult to spot the LS. Daffers was looking at the navigation computer and was talking me down.

'Zero Nine Seven degrees, nine miles . . . Zero Nine Four degrees, eight miles.'

While he was looking at the instruments I was looking around. In a split second I could see the signal, a sharp beam piercing the sky. The AH had cleared us.

'Doorman 26, Wildman – MacAndrew,' said McGinty over

the radio, against the cacophony of the Apache's 30mm cannons spitting shells at the ground.

Even though I could hear him shooting in the background, I barely registered it as I had invested all my energy in bringing the cab safely down and I was now in this zone. We had all the firepower on our side and I had to trust that it wasn't our time. Our destiny was in my hands – it depended on me not porking up the landing – but also in the hands of the guys around us, pounding the enemy to get us in.

There was an increased tension. I could almost feel it emitting from the headphones but at the same time we all just wanted to get the job done and then get out of there. We had come here to pick up the casualties, to get the guys out, and that is what we were fucking well going to do. Failure was just not an option and I had no intention of being the first Brit Chinook to be shot down by the Taliban. No way!

I knew that there were people protecting the landing site on the perimeter as well so the whole place had been malleted. As we came into the landing site it was like we were descending in a slow-motion, ethereal manner and bizarrely there was no tracer and no small-arms fire.

'One minute, guys, one mile to go, Mark – reference the large burning building in your two o'clock, the HLS is just left of it now. Half a mile to go, HLS on the nose, 1k, just right of the nose behind the smoke cloud, 500 metres now – you good visual, Mark? There it is, twelve o'clock low, good speed, good rate of descent. Everyone secure?' said Daffers.

As we came into land, we set up into the gate at 100ft, 30 knots. Daffers was talking to me down.

'You've got the site. Seventy feet, thirty knots.'

'Little farm, fifty feet, twenty-five knots.'

I brought the nose of the aircraft up and pulled the power in. The crewmen took over the talk-down.

As we descended on a running landing Sam, Dan and Spence were chucking medical supplies out of the windows and off the ramp before the wheels even touched the ground. The dust cloud started rising, mixed up with straw swirling around. This just emphasized that we were flying into a feudal, backwater hellhole.

The back wheels hit first. Then the front.

As soon as I felt the rear wheels touch, in my head I repeated a mantra: *Keep it straight, keep it straight.*

Those last seconds were crucial. I couldn't see Jackshit and I am relying on the Force and my Spidey sense to deliver a decent landing. All my instincts were working at 300 per cent. It was about striking a fine balance between a solid landing and taking the wheels off. There couldn't be any lateral movement or the wheels would be toast. Daffers leant across and tapped my arm: 'Good effort, mate, good landing.'

I nodded.

'Parking brake on, clear ramp,' I said.

'Ramps travelling,' said Sam.

'Casualties are coming out.'

Daffers and I were doing all the checks to get out.

'Brakes on the toes, AFCS is in, the RADALTs are bugged and the kit is all armed and ready – you good to go in the back?'

Daffers said: 'The quickest way to get out of here is to come up, turn right, come into the wadi and get the fuck out of Dodge, but it's a bit obvious.'

We looked at each and said together: 'Let's go left.'

As they were doing the handover between the medic and the MERT, Dan piped up on the intercom. 'Which way do you want to go out, mate?' he asked. 'Are you going to go over the top, out to the south-east?' he said.

I looked up and there was a building ahead of us. It wasn't

very high. About one and a half storeys. It was the logical exit but for some reason it didn't feel like the right one to me.

I replied: 'Nah, I don't want to go that way. It seems too fucking obvious. We'll come back over and we'll come back out over my left shoulder.'

'Ok, roger that,' he said.

The casualties are onboard about forty-five seconds later and the ramp is lifted.

'Ramp up,' said Sam.

'Pre-take-offs done, lifting,' I said over the intercom.

'Roger, clear, above and behind,' said Sam.

I pulled power and came into the hover. At the same time I was also pulling back stick and power, transitioning them backwards. I kicked to the left and pushed the pedal in. The aircraft turned about its centre and did a J-turn. It's called an over-shoulder departure. It was the most natural thing to do. I could hear the gearboxes and engines screaming when I flew the aircraft like this, the stench of hydraulic fluid, gunpowder and sweat filled the aircraft, the dust raised by the take-off was thick and coated my mouth, dry with the stress of what we were doing. As soon as I backed off the power and levelled off to build speed, the aircraft stopped screaming and started to accelerate away fast.

We came back out and we were going to boot out at low level fast. North of the District Centre, across the wadi and then start climbing rapidly to the safety of height.

We were mega exposed but as soon as we pulled up, went back and pulled the aircraft round, the Taliban fired two RPGs from the one o'clock, location of the medium-height building I'd seen earlier. The AH swooped and banged out a Hellfire missile totally annihilating the building. If we hadn't made that snap decision to take the exposed route we would have been dead and those RPGs would have shot us down for sure. As we lifted up at the four o'clock some Taliban dude opened fire. It was a big tracer so it was likely to be a .50 cal or a 12.7mm Dushka, a big Russian-made anti-aircraft gun. He was tracking us and it looked like it could hit the aircraft.

Dan had a huge flashback of when an RPG in Iraq had hit the side of the cab they were in and left them with eight casualties onboard. He opened fire and yelled: 'BREAK RIGHT!'

We broke over it. Sam and Dan opened fire and so did the guys on the ground. Triple whammy. He got taken out from three angles. The enemy's heads were down from getting schwacked by the Apaches and they had malleted the fuck out of the area.

As we broke right Dan opened fire. He had cases and cases ricocheting off the inside of his catchbag going backwards. One burnt him on the inside of the arm, a couple went backwards and hit the back of Spence's thigh. As Spence looked down he saw the tracer from the RPG whiz past the aircraft. In that split-second he thought he had been shot in the leg. At the same time Daffers saw gunfire speed beneath his left foot. In that moment Sam and Spence collectively opened fire and began raining bullets down on the insurgents below. The Taliban responded in kind and the skies were amok with the chaos of crossfire.

I looked up and saw tracer heading directly for the aircraft. Over the intercom, I squeaked: 'Tracer' – for some reason it came out in a high-pitched falsetto tone. I cleared my throat and said again much more deeply: 'Ahem, tracer!' Even in the heat of battle I knew that squeak was coming back to haunt me once we were back at Bastion. They were not going to forget it in a hurry. I don't know, you shag one sheep!

If we had turned right they both would have hit us and we would have gone, the crew would have gone, the doctors would have gone, the casualties. It would have been catastrophic for the families, the squadron, the station, the media – not to mention the political implications. The domino effect was unthinkable. Not that we cared too much about the political fall-out but the repercussions across the scale would have been immense. We were literally dodging bullets as I threw the cab around at speed like a 100ft bucking bronco, ascending at 120 knots into the celestial safety of height.

Beneath us Dan could see a Taliban gunman tracking us. Dan turned the M60 on the gunman and took him out – dead. His line of sight was directly on the aircraft. If he had fired it could have taken me or Daffers out, or the engine and transmission. Boom, aircraft crashed followed by enormous shit magnet of trouble. Nobody wanted to lose a Chinook. There are another twelve blokes on a Chinook. Imagine what happens when a Chinook gets shot down. We send another one. It becomes a Chinook pile-up. Imagine how many casualties would be on the ground. Is the area secure? How are we going to secure it? Are there enough forces there to pick up the casualties? Potentially it becomes an enormous op. Dropping off more people are we going to secure a landing site? Will troops have to fight into the town to secure a landing site to pick up casualties? Bad juju.

Meanwhile back at Bastion, Giles's crew had been stood up. They were sat on the pan turning and burning as an auxiliary measure just in case the whole thing went tits-up to pick up the pieces. In the back of the cab they are dealing with the casualties.

The doc came on the intercom: 'The casualty is stabilized. You can take it steady. One of the guys has a neck injury so as steady as you can, mate.'

'Thanks, mate,' I said. We headed back at a comfortable 120 knots.

Once we hit height it went quiet on the intercom. The adrenaline dissipated from our bodies. We had done it but now we just wanted to get home and go to bed.

The Apache pair that had flown the most hours were left to escort us out but McGinty and his pair stayed on until they were running out of gas. They had just stirred up a hornets' nest and 3 Para were getting a pasting. Rockets were flying everywhere and what we didn't want to do was create another scenario where there was another casualty and they would need to be CASEVACed out.

We landed on back at Bastion at the hospital HLS. There was an ambulance waiting to take the guys to the hospital.

Colonel Tootal and the RSM were there. The MERT took the guys to the hospital. We repositioned to our normal spots and the rest disembarked. The protection team got off and all said: 'Cheers.'

As we shut down the cab, it felt like I was shutting down as well as the adrenaline seeped out of my body. We had done it. We had picked up the casualties, I hadn't porked the landing and we hadn't been shot down. We had all lived to fight another day.

Daffers patted me on the arm. 'We didn't get them earlier but then we went back and we did get them.'

It was important for the guys on the ground to know that airforce guys won't shy away from it when the shit hits the fan, that we'll come and get them. This was why we did it. We did it for them. The guys on the ground. If we didn't do it I don't think I could look at my face when I was shaving in the morning. Those guys rely us on so we are going to go and do it. It was a really great feeling to know that we had got them.

I said to the guys over the intercom: 'Well done, lads. We got it done. We got the job done.'

Immediately at that point an unbreakable bond was

formed. We all had this knowledge that we had just achieved something. It was a knowledge that we had all shared. It was a night that none of us would ever forget but something that the only the five of us would understand. It was a private, personal moment that bonded us as five blokes who would be mates for life.

Nobody else will ever understand what it was like to go back into Musa Qala on that night, under those circumstances. Only those who do something that is equally dangerous will understand the special bond that exists between us now as a crew.

It is going in and doing something and then a few hours later going back into the same place where you have just had your arse handed to you. That is a whole different kettle of fish. You can't explain to the outsider how you feel but among the crew you know, and that is what creates the bond. I have more of a bond with these guys than any other crew that I have ever flown with because we all got on but more importantly we did the big thing together. It was a team effort and there was no one member more deserving than another. It was all or nothing.

We headed back into the JHF(A) CP to debrief. We were still pumped with adrenaline and felt edgy.

'Come and have a look at the gun tape. You have been lucky as fuck tonight. How those RPGs missed you I will never know.'

Watching the footage from the Apache gun tape, we saw how close to death we had come on the initial run into Musa Qala. As we continued watching the tape we saw how on the way out, as we turned left, two bright streaks whizzed past the right-hand side of the aircraft. Just watching it made me throw my head back a little. It was that close. The RPGs had missed us by a matter of inches.

'Fucking hell!' I was stunned. I was watching certain death pass us all by a cat's whisker.

Daffers and I looked at each other: 'Why did we go left?' I said. 'Why did we go left?'

'It put us more over the town, which gave them more of a chance to shoot at us. What made us go left? When the most obvious thing for us to do was to go right.'

'That Teletubby shooter was sat there, looking at us and thinking they are going to go right because it is the most obvious exit. He must have deliberately aimed to catch us on the right-hand side. Ha, ha. Got you, sucker,' laughed Daffers.

In many ways we had to take our hats off to the Taliban because we had burnt them, we had bombed them, we had mortared them and we were shooting at them and yet they still waited to get us on the way out. The Taliban were not cowards – no matter what we threw at them they were prepared to have a go.

I took the news with mixed emotions. Part of me was pleased to have made the call, which effectively saved all our lives, and part of me felt sick to think that if we had elected to take the obvious route we would now all be dead.

'We didn't get the Spectre C130,' said McGinty.

'There wasn't one available. It was the Hog that ripped the place to pieces.'

We discussed how we had come up with the plan and how the plan had worked. It was agreed that flattening the place on entry had given us the respite we need to get the cab in to land.

What we hadn't realized was that, on lifting, all hell had been unleashed and that the AH were firing flat out from the moment we became airborne again.

'I fired three Hellfires,' said McGinty. 'The minute you took off, the Taliban opened up with everything they had. They used mortars, RPGs and small-arms fire. Both of the AHs didn't stop firing from the minute you took off until we were out of fuel. 3 Para gave it everything as well. It was a dogfight.'

We would have handed the Taliban a huge physical and moral victory if they had taken a cab and it was a fucking miracle that we hadn't. At the time I hadn't realized how close to death we had come. I left the CP tent and I reached for the Reds. As I lit the cigarette and chugged on it I could hardly taste it. I was smoking but barely aware of it. I walked back to the IRT tent with Mike McGinty. We were both completely knackered. We had exhausted every ounce of emotional reserve and with every step back to the tent the energy was seeping out of my body as I gradually came down off the adrenaline-induced high.

'What a fucking day,' said McGinty.

'I don't need another day like that in my lifetime.'

'What I could do with now is a cold beer.'

'Yeah, there is nothing like a chilled beer to help you unwind after a hard day at the office,' I said, half smiling to myself, just at the thought of it. Right then, I could have murdered one.

Mike broke off and headed over to his tent. I walked over to the IRT tent. Sam was outside smoking a tab. Dan and Daffers were there too. I reached in my pocket and pulled out a Red. I joined up with them. 'Nothing wrong with a bit of chain-smoking after a day like today,' I quipped.

It was about 1 a.m. All of us were ball-bagged and we just wanted to get our heads down.

'I am off to KAF first thing, early doors,' said Sam.

'Right, mate. In that case, well, lads, I just wanted to say what a fucking hoofing effort that was on everybody's part. What we did tonight was above and beyond but only we'll know it. Don't expect anyone else to understand because it's likely they

Dave Westly and I in a tent in Norway in 1995. Dave would be flying my AH escort into Musa Qala some eleven years later.

On deck aboard the USS *Essex* during my time on exchange with US Marine Corps flying AH-1W Cobra Attack Helicopter. The nose of one of the Cobras is visible behind me. So too is the three-barrelled M197 20mm cannon.

In Helmand. Yours truly standing in front of the office. The M134 six-barrel Minigun is visible in the side-door. We had two.

Home Sweet Home. The grots at KAF.

Watching TV in the IRT tent in October 2006. The home comforts would improve rapidly over the coming years.

Refuelling the beast at Camp Bastion.

Looking down the line outside the IRT tent at Camp Bastion.

Camp Bastion, early 2006, prior to its massive expansion. Note the size of the dust cloud indicating a Chinook has just landed.

A Chinook landed on in the Green Zone on an IRT shout. One of the most vulnerable times for the cab and crew.

The 'bird strike' aka a large-calibre bullet which clipped the rotor blade on the way into Musa Qala.

The business end. The view from the cockpit while transiting over the Red Desert.

The views could be incredible. This is Kajaki Lake from height.

The clear delineation of the edge of the Red Desert.

Outside temperature gauge indicating a balmy 47.3 degrees. The air-conditioning consisted of opening the window!

The ever happy Daffers with 18 Squadron 'B' Flight pilot Russ 'Icon' Norman.

Me with one of the crews for 'the obligatory crew shot'. Typically, the only crew shot that I don't have is one of Sam, Dan and Daffers and myself!

The mighty Chinook doing what it does best, carrying a 4-tonne load at high altitude.

Smoke laid to provide an accurate landing position in the final stages of approach into Now Zad to re-supply.

A fantastic shot of a three ship conducting a pick up in the Afghan mountains. This picture now adorns many a Chinook pilot's living-room wall.

Troops waiting to load aboard two Chinooks for the Sangin RIP as seen through NVGs.

Ready to go. The apparent calm belies the complexity and intensity of what lies ahead.

Lifting out a 105mm Light Gun for the MOG.

Back of a cab loaded with 2 tonnes of ammunition.

Loading the cab for framework tasking. We support all nationalities of ISAF troops in theatre.

Back of cab, not even half-full, during framework tasking.

Four-tonne underslung load viewed through the belly of the aircraft.

Positioning to pick up an underslung load in the GAFA.

The final stages of an underslung load drop-off with a rising dust cloud about to envelope the cab.

A Chinook landed on in the GAFA waiting to load.

Browning out prior to lifting after troop drop-off down near Garmsir.

Routing out of Bastion at dusk. You never tire of the amazing Afghan sunsets.

won't. Has anyone got anything they want to add before we get our heads down?'

Nobody had anything to say. Everyone just wanted their bed.

'Dave and his crew are picking up the IRT tomorrow a.m. so get your heads down and we'll pick it up slow time tomorrow. Sam, mate, I'll see you when you get back in a couple of days,' I said.

As we walked into the IRT tent, Dave and his crew were still asleep. Blissfully unaware of the shitfight we had all been through. Very few people knew about it. Colonel Felton had been watching the MIRC chat back at KAF but most would have been asleep. From the Chinook Flight it was only the five of us, and Giles's crew as well as they had been turning and burning on standby. We were amazed that so much had happened in one day.

'Should I feel bad because I have shot dead at least six or seven people? And yet, I don't feel bad. Should I feel bad that I don't feel bad?' said Dan.

'It was either them or us,' said Sam.

Dan had a little prayer for their heathen souls. He prayed for them to rest in peace. 'I didn't know them but they could have been good people, fathers and sons.'

Sam was right, it was either them or us, kill or be killed, and we had fought harder so it was them.

'I feel like I should feel guilty but I don't,' he said. It was a conditioned response – if someone is shooting at you, you shoot back.

Until our time in Afghanistan air-to-ground gunnery was paid lip service. If the crewman fired even one round, there was usually a stoppage straight after and that was it. So it was significant to be in theatre hosing down 500 rounds. It wasn't normal practice for the operational Chinook squadrons. This war wasn't like any that had been fought in the history of the Chinook's workload for the British armed forces in any theatre.

The campaign had a huge impact on some of the guys, particularly on the crewmen. It was because of the stark contrasts. One minute they were watching comedy on television and the next minute they were picking up body parts.

Similarly, in the late Bosnia, early Kosovo, campaign, when the Tornado Wing launched from Brüggen to do operational missions in Bosnia the pilots and aircrew experienced post-traumatic stresses related to extreme contrasts. They took off at midday, flew all the way around France to Bosnia, did the bombing and then returned back to the base in Germany. They started their day at 0700, said goodbye to the wife and kids, went to work, loaded their weapons, armed the aircraft, took off, flew to Bosnia, bombed the shit out of it, flew back, landed on and went home for tea. Very bizarre circumstances not dissimilar to the Battle of Britain, where the crews were scrambled, flew a mission and returned just hours later with one crew down if they were lucky. It is the speed at which people can be extracted from the most horrific and traumatic of circumstances and sent back to the relative normality of their domestic life that has a huge psychological impact on them.

We were the last aircraft into Musa Qala. No aircraft flew back there after the 6th because tactically it was a dreadful place to go into and it was becoming so likely that we would lose an aircraft. We tiptoed to our beds. The tiredness ebbed through my body as I got undressed. I climbed in on my last legs, shut my eyes and slept the sleep of the dead.

In the morning I was awoken by Dave and his crew as they were getting up. They were the IRT because they were the only ones on crew duty. We were on standby.

'Everything all right last night?' asked Dave.

'Yeah, but we took a couple of rounds,' I replied sleepily.

'Shit! I hope no one has a pop at us today.'

'Yeah, right, it's always better if they don't, mate,' I nonchalantly replied.

I sat up and rubbed my face. I was still pretty knackered and yet had a raging attack of the munchies.

'Scran anyone?' I said.

'I am in,' said Dan

'Yep, me too,' said Daffers.

Sam had already got up and had left for KAF to do some tasking. We got up, threw on some combats and headed over to breakfast. We scoffed down some grub and then headed straight back to the tent for some more shut-eye. We had nothing planned for today and there were two crews between us and any action so we were chilled out! I did, however, want to go and see the tea-strainer of a cab we had left with the engineers after our first trip into Musa Qala.

Chapter 22

After lunch, Daffers, Dan and I got up and out of the tent. Finally we felt more human. 'I think we should go and check out the cab that went U/S. The Line told me it had taken rounds,' said Dan.

We jumped in a Landie and drove over to the LS.

'Hello, mate, we are looking for Wilf,' I said. 'Is he around? We want to check out the cab that went tits last night.'

'He's over there, sir. ' The young engineer pointed to a group of engineers who were stood next to one of the aircraft. We walked over to them.

'Hi, mate, I was flying the cab that went U/S last night. Apparently it took some rounds,' I said.

'Took some rounds? You are all fucking lucky to be alive. Come and look at this.'

The aircraft had taken four bullets. One had gone through one of the rotor blades at the back, another above Daffer's head in the bulkhead. The most significant hit was the bullet hole in the blade root. The bullet had pierced the main part of the blade an inch away from potentially taking the rotor head off. If one blade had parted from the head the imbalance created in the other five blades would have resulted in a catastrophic desynchronization of the front and aft blades. This would have guaranteed certain and instant death for all the crew and passengers.

Chinooks have a reputation for being resilient and surviving incidents against all odds. The most resilient of all the CH-47s was Bravo November, serial number ZA718 (Boeing

162

construction number B-849). This notorious Chinook has in fact seen action in every major operation the RAF has been deployed to in the helicopter's 25-year service life and is still in service today.

Bravo November was part of a Chinook four-ship that was supposed to deploy to the Falkland Islands. The *Atlantic Conveyor*, the container ship carrying them south, was sunk by an Exocet sea-skimming missile on 25 May 1982 by an Argentine Navy Dassault Super Étendard. While the other three aircraft sank along with the ship, Bravo November was airborne at the time, picking up freight from HMS *Glasgow*. It then flew to the safety of the aircraft carrier HMS *Hermes*, where it was nicknamed 'The Survivor'.

Later in the conflict, during a night mission, Bravo November was transporting three 105mm guns to British troops when the pilot, Squadron Leader Dick Langworthy, a thick snow shower clouding his vision, collided with the sea at around 100 knots (175 km/h). The collision was caused by a faulty altimeter, and the impact with the water flooded the engine intakes. By some miracle Langworthy and his co-pilot were able to get the helicopter airborne once more. Unable to navigate, with the radio damaged, Bravo November returned to San Carlos to discover the impact had caused nothing more than damage to the radio systems and dents to the fuselage. Once again living up to its reputation as the Survivor.

The redundancy of the Chinook has got us out of a lot of problems such as knocking wheels off and still being there, losing entire fuel systems and still being there. Long may it continue. Our good fortune was not lost on either myself or Daffers.

'That, my friend, is our bird strike.' We both looked at each other and laughed. 'Mate,' he said, 'I am not kidding. I really thought we'd hit a bird.'

It seemed funny there and then looking at the cab. Even at the time, in the height of the contact, we didn't think that we had come an inch away from desynching the rotor heads. We actually thought we had hit a bird. It was just a single dark object. Potentially it was a single round from a rifle, in which case it was a really good shot. For a marksman to hit a moving aircraft head on was either a lucky shot or a really good one.

The reason that we have not lost an aircraft in Afghanistan is because we have been lucky. If they took a bit more time to tee us up then they would make the shot. It was a combination of luck and the fact that the Taliban were rushing to take the shots. The Taliban knew if they hung around too much, either the Chinooks were going to fire at them or the Apaches were going to waste them.

They would fight and then leg it. The Taliban might want to fight for their cause but like everyone they would rather not die. We were either in the right place at the right time, the right place at the wrong time, the wrong place at the right time or the wrong place at the wrong time – you can take your pick, but for whatever reason it wasn't our time to die.

How we had responded in this situation was a product of our training and conditioning. It had served us well. None of us had ever sustained such heavy fire but we all kept calm and focused on the task at hand.

Even in the height of all the tension we were always smiling and laughing. It is only afterwards, maybe a few days later, that the severity of what has happened and the real implications dawn on you, but then you push them to the back of your mind. There is no point dwelling on what could have been but didn't happen. It reinforces how fucking lucky we were.

Chapter 23

There was a crew change, a frequent and common occurrence. Tweedle Dee, a very professional aviator and good egg was now my co-pilot. Sam had come back from KAF and was on the crew again. I had now established a system whereby every crew Forward rotated every twelve hours. It worked out as twelve hours' IRT, then twelve hours' standby and twelve hours' HRF and so on.

Things had quietened down a bit for us and we were all glad of the downtime. We couldn't have continued at that pace without us all burning out. Burnt-out pilots and aircrew start making mistakes and that's when it all becomes even more dangerous because that's when accidents start to happen.

We could never truly relax because we never knew when the next shout was coming. There was a bit of hanging around, playing computer games, watching *Firefly*, endless Marry, Shag, Kill. Sam, however, was practising the art of cock lunging. And, in the process, winding Dan up. Lunging is one of the finest crewman sports ever invented. Pilots occasionally participate but it's the crewmen that have made it into a fine art. The art to a basic lunge is to slot the lunge in quietly but unexpectedly. Advanced lunging involves placing the forward foot on a higher point, such as a chair, table or aircraft step.

Cock lunging takes the ingredients of the advanced lunge and adds an inappropriate twist, thus taking the lunger and the lunging target to a new level. The target is a poor, unsuspecting sod sat quietly in a chair, in this case Dan, when along comes the cock lunger, in this case Sam, up comes the front leg and in goes the lunge. The lunger's crotch is presented dangerously close to the target's face – hence the term the cock lunge.

He tried to do it to me once.

'Ha ha – Sambo Solo strikes again, stealthy but armed and dangerous! Unbeatable.'

'Aargh! Get your sweaty nut sack out of my face, arsewipe!' I goosed his nuts.

'I am bored,' said Sam.

'I bet that I know more ways to say pussy then any of you tossers.'

'Bring it on,' said Dan,

'I'm in. I see you a minge, fanny, twat,' I said.

'I raise you a beef curtains, axe wound, doner kebab, love tunnel, shaven haven, muffin hole, gash, man trap, box, hairy muff,' laughed Dan.

'Badger's den, wizard's sleeve, otter's pocket, hairy goblet, the end zone, love tunnel, horse collar, busted sofa end, pee pee flaps, vertical ham sandwich, hangar doors, punani!' smirked Sam.

'Otter's pocket! That is brilliant,' I laughed. 'Hoofing. Err . . . flange, bearded oyster, furry cup, badly packed kebab . . .'

'Did you hear that they have been playing top ten male shags on 27?' said Dan.

'That is so wrong,' I said. 'Not afraid of trying it but just afraid of liking it.' We all laughed.

Boredom fuelled it. We just tried to gross each other out. We would try to think of the most disgusting thing we could to take our minds off the endless waiting around or long transits at height where the flying is pretty dull. Games such as what is the most disgusting thing you would do for a million pounds. It keeps your mind off the fifty million other things you would rather be doing than being sat in a tent in the middle of a shithole waiting for something to happen.

Dave Westley stuck his head in the tent: 'Mark, can you head over to the CP, there's a task come up,' he said.

'Yeah, sure thing, mate.'

I sat up from my cot and reached for my boots, put them on and laced them up. I grabbed my pistol, stuck the Reds in my pocket and walked over to the CP.

'Morning, gents,' I said.

The lads in the room gave me a wave, some not even raising their heads. OC 'B', who was the programmer, called me over. 'We've got some tasking over at Sangin.'

There was a load of resupply troops, ammunition, food and water that needed to be flown in.

'You have also got Bill Neeley and an ITN crew that will be embedded in Sangin for a couple of days,' he said. 'This will be briefed into tomorrow's a.m. brief and then will fly out the next morning at 0400.'

Bill Neeley was one of the first broadcast journalists to be given access to the frontline. It was a bold move on his part. Having a journalist in Sangin would give the British people access to news about the conflict from the frontline for the first time.

At this time the apathy of the British people towards the conflict annoyed me. The constant media and public obsession with celebrities like Posh 'n' Becks when our lads were back in this shithole, fighting for their country, being bombarded by a barrage of constant enemy fire, just drove me apoplectic with rage. This war is about keeping Britain safe from terrorists. Terrorists train in Afghanistan. The Taliban protected Osama Bin Laden and refused to hand him over to the allied forces. There will be terrorist attacks on Britain. This conflict, which is being played out thousands of miles away from the UK, is taking the lives of our sons and daughters and at the time all the papers were interested in were the lives of stupid, gormless chavs who have a high profile but contribute nothing. Seriously, who gives a fuck?

I am afraid this tended to spill over into my view of journalists. On the one hand we needed the media because we wanted the story to be told. However, at the same time we needed the media to be mindful of the fact that our jobs aren't just a game, that they could dip in and out of them as they see fit. I hoped Neeley wouldn't disappoint.

As Sangin was also being pounded on a regular basis by the Taliban, to get the resupply and Bill Neeley in we had to co-ordinate the exercise as a deliberate op. We were needed to fly it under the cover of darkness and it required some extensive planning. The lift was planned for 0400. To get our crew duty hours in line we had to be in bed at 1700 so that we could get enough sleep in.

On the morning of the resupply my alarm went off at 0100. It was dark and cool. It had been planned as a two-ship deliberate operation. Tweedle Dee, Dan and Sam were on my crew. We all rolled out of our pits, scratched our arses and barely uttered a word as we mustered our shit together.

Sam and I met outside the IRT tent and had a smoke. This first morning Red tasted like shit. Sometimes I wondered why I bothered but I soldiered through it and washed it down with two lukewarm cups of coffee. It was a bumpy jump-start to the day. The mess weren't cooking breakfast for the aircrews, which was definitely a flaw in the planning. No food in my stomach meant that breakfast consisted of as much caffeine and Reds as time allowed before lifting.

We all jumped in the Landie and headed to the cab. We arrived and the troops were waiting round the back. I walked straight up to the cockpit to begin the pre-start-up checks.

Sam and Dan made sure the troops, supplies and the underslung load were ready to lift. With the engines running we were sat waiting in the cab to get clearance to depart, the rotors turning, when we got a call over the radio saying the AH had gone U/S and that we were aborting the op today.

'Fucking hoofin!' I said. 'Sam, Dan, did you hear that? We are not going, the AH is fucked.'

'Brilliant,' said Sam.

'Roger,' said Dan.

'Can you let the troops know and unpack the aircraft, mate?'

'Yep, roger that,' said Dan and Sam together.

'At least we can go and get some breakfast now,' said Sam.

'Sam – can you tell the guys at the back, same time and place tomorrow,' I said over the intercom.

'Roger.'

I looked over at Tweedle Dee.

'I bet that went down like a pint of warm sick,' I said.

We both chuckled. Not that it was any better for us. None of us really appreciated a 4 a.m. lift. All of us, including the troops, would have the whole day hanging around. For us it would be back to bed at 5 p.m. to get enough sleep to keep us in crew duty.

As aviators we are bound by many different rules and regulations. One set of rules that if mentioned to ground troops is guaranteed to piss them off is aircrew crew duty. In a 24-hour period, we are allowed to spend a maximum of sixteen out of bed and then we must get eight hours of uninterrupted sleep. This sleep period is essential to keep us and our passengers alive. Think of what you are like driving a car after sixteen hours, if you have ever done that, then magnify it with the stress, noise and vibration we are subjected to when we go flying on operations in the Chinook. Most troops understand that after we explain it to them. The rest just think we are a bunch of prima donna twats.

At least it gave us an opportunity to try and arrange some breakfast for tomorrow.

We shut down the cab and jumped into the Landie. As we were driving back to the IRT tent we all agreed that we were starving. Dan was nominated to go to the canteen and rustle up a round of butties to bring back to the tent before we all got our heads down. He dropped me and Steve at the CP so we could hand back the Green Brain and confirm that we were going to repeat the same thing tomorrow.

McGinty had just come into the CP to start his shift.

'Yep, the Apache is tits but the Line think they will have it back in service for a 0400 lift tomorrow, lads. No change to the plan. Back on HLS at o'dark early tomorrow then.'

As we knew from recent events the HLS in Sangin was a field, which was relatively exposed and surrounded by Taliban. There were few options into the approach and so the Chinooks were incredibly vulnerable. There was always a massive likelihood of taking rounds. Since Musa Qala the tension had increased tenfold and the Taliban attacks had become more frequent and intense.

The Taliban were angry about Musa Qala so they had an even bigger axe to grind. The significance of taking down a Chinook had amplified. The risk involved in the resupply was enormous. I was stood outside the IRT tent when Daffers walked by. He looked a bit pissed. I stopped him.

'Are you all right, mate?' I asked.

'I am having a few issues with my other seat,' he said. He was talking about the other aircraft pilot.

'He's got the fear. I can see it in his face. He's not said anything to me yet and I have not said anything to him. But I think it's affecting his decision-making, which means his operating is compromised. He's making me nervous because he's handbagging. At the moment he's likely to get us killed before the Taliban fucking shoot us down. He's just being a bit woolly, y'know. You should have seen the relief on his face when we were stood down. He became a different person.'

This guy was young and he had young kids. The threat was high. The word was out that we had the shit shot at us in Musa Qala so it wasn't unreasonable to be frightened. It was a difficult one. What do you do? He had two choices – he had to man up or step down but who was going to say it to him? And it's a difficult thing to do, step down and say, I can't do this because I am worried about dying. If you step down somebody else has got to step up. It's not the team way but at the same time dithering around like an old lady and faffing about instead of making confident decisions didn't do anyone any good either. It had to be sorted out.

Daffers by the look on his face was pretty close to gripping it. At the end of the day, Daffers had a wife and kids too and he wouldn't be prepared to die because some young kid had lost his bottle but wasn't prepared to admit it.

'What are you going to do? Have you thought about it from his side? He's not been into Sangin or Musa Qala yet,' I asked.

'Valid point. I dunno yet. I am going to give it some time,' he replied thoughtfully and walked off.

I chucked my tab end in the bucket and headed back inside the tent.

Chapter 24

On the *second* morning of the resupply my alarm went off at 0100 again. We got up, smoked and drank coffee. Same shit, different day.

Sam and I met outside the IRT tent and had a smoke together. This first morning Red tasted like shit. Sometimes I wondered why I bothered but I soldiered through it and washed it down with two lukewarm cups of coffee.

A few minutes later we started the engines and heard the whump as the rotors slowly gyrated into life.

Over the radio the Tower gave us the all-clear to go.

When we lift to the hover with a load hooked underneath we listen to what the crewmen are saying very carefully. It's called voice marshalling and crewmen do it very well. Basically they are telling me what to do with the aircraft.

'Up gently and left,' came the command from Sam. 'Up five and left . . . Load coming clear, up two only. . . Steady, height is good, good load, clear above and behind.'

The last bit let me know that the load was ok and that we were clear above and behind the aircraft to move from the hover into forward flight.

We were about halfway to Sangin when we got a call over the radio saying the AH had gone U/S and it was a mission scrub. Once again we were aborting the op today.

'Fucking hoofin!' I said.

'Sam, Dan, did you hear that? We are not going, the AH is fucked again?'

'Brilliant,' said Sam.

'Roger,' said Dan.

'Can you let the troops know that we are going back?'

'Yep, roger that,' said Dan and Sam together.

'I am fucking starving,' said Sam. He didn't handle early starts without scran very well.

'Sam – can you tell the guys at the back, same time and place tomorrow,' I said over the intercom liked a scratched record.

It felt like groundhog day. I couldn't wait to get out of this sleep pattern. Mind you, I wasn't going anywhere without AH support so until the cabs were fixed I had no problem going back.

Clearly the tension with the Taliban was ramped in theatre and we had heard about it back in the UK on the bush telegraph and then experienced it first hand once we started operating. However, in Blighty at this time they were still more interested in mindless celebrity gossip. There hadn't been that many journalists at the frontline genuinely experiencing what the lads were going through, so nobody really knew what was going on in the desert thousands of miles away from the motherland.

On the *third* morning of the resupply my alarm went off at 0100. You know the drill. This time I hoped we'd get airborne.

The cab was full to the gunwales. Thirty troops with kit and weapons, crates of ammunition, food and water and a load underslung. It had become pretty clear that getting supplies to Sangin wasn't a guarantee so they made sure that we were packed to bursting. In order to get Bill Neeley there as safely as possible it had been decided that we would use the HLS outside the DC. This was partly to mix it up a bit and avoid predictability and also to ensure that the HLS was protected and surrounded by ground troops. Several dozen soldiers and a Scimitar light tank had been sent out to secure the desert landing area.

Third time lucky! We lifted, attached the load and headed out towards Sangin. As we approached the landing zone, the crewmen test-fired the guns. We came into the gate and brought the cab down.

As the dust cloud formed, Dan was counting me down and watching the ground so he could release the load.

'Load gone,' he called over the radio.

Once I heard the load was released I concentrated on bringing the cab forward and dropping the rear down so that the wheels touched.

'Ramp down,' said Sam.

Sangin was attacked every day so as soon as the ramp touched the ground the troops, Bill Neeley and his cameraman bundled off the back of the aircraft. There was a sustained burst of small-arms fire, and an RPG flew over their heads at a height of around twelve feet. Then another bounced off the ground close to them and landed in a nearby river. The British troops that disembarked the cab and those who had secured the landing area fired back.

Around a dozen Afghan soldiers fired more weapons and RPGs. The firing was intense. The Apaches circled overhead to neutralize the targets. They began to shoot at the Taliban, firing sustained bursts into the cover where the Taliban were shooting. They were backed up by the base at Sangin and the troops fired 81mm mortars.

Neeley and his cameraman reported back. 'It was very hairy,' he said to the camera. 'Especially when I saw an RPG coming towards me, but I didn't feel that at any time the British troops were not on top of the situation. Everyone got back to the base safely. In fact, there wasn't a scratch on anyone. I asked one of the troops if he had been surprised by the attack. He told me: "When it comes, it's still a shock but not a surprise."'

His cameraman was quoted as saying: 'We had just landed and were physically manhandling the equipment on to vehicles when we came under heavy attack from rocket-propelled grenades and gunfire – at which point all hell broke loose. The sky was lit up with tracers – both coming our way and going their way.

'We were under the wing of the officers in charge, who took complete control of the situation – securing safe passage into the base.'

Three days later Neeley broadcast the first TV report from the frontline back to Great Britain. It was a turning point in the conflict because for the first time back home in Britain the British people were seeing that this was a war.

Bill Neeley had been ambushed by the Taliban. He was the first to tell it and show it how it was. It needed doing and he was brave enough to do it. He was not the enemy within because he saw for himself what the troops were enduring and he respected their bravery. It was reassuring to know that there are some journalists out there who let the truth be the great story.

Chapter 25

I was ready to get back to KAF for a few days. Apart from anything else I was bingo clean pants and I had barely had a conversation with the DCOS since I had arrived here. It would be good to not be working in an environment that could go from 0 to 60 in a matter of minutes. It was exhausting.

But there was one last task between us and a return to the Green Bean and the launderette. We had to lift an Explosive Ordnance Disposal vehicle out of Sangin. One of the troops had inadvertently left Sangin with the key to it in his pocket so the vehicle could not be moved. The EOD vehicle was needed for another job elsewhere in the province.

The vehicle looked like a *Banana Splits* buggy, but it was a useful piece of kit in a counter-insurgency campaign.

It took two attempts to extract the EOD. First of all, it was decided that the simplest solution was to drop the key to them, by placing it in an ammo box and then dropping the ammo box out of the window on a regular resupply task. That sounded like a plan, and anything that avoided having our arses handed to us in Sangin got my vote. That is what we did.

However, they realized that even with the key they couldn't start the vehicle because it was shagged so we were stood up again to go into Sangin to pick up it up and bring it out underslung at night. The problem we faced was that we needed to get in and out of Sangin quickly, but picking up an underslung was not normally a quick evolution.

It was all change again crew wise. Now Dan, Sam, Daffers

and I were on shift as HRF and we sought the advice of Mr Mack, the master aircrewman on the flight, who was an incredibly experienced operator. Mr Mack and I were scratching out a plan in JHF(A).

'I don't want to be hanging around for any longer than thirty seconds,' I said. 'What's the best way of skinning this cat?'

He thought about it. 'You don't want to be doing an OCF load.' The OCF – Operational Conversion Flight – is where baby Chinook pilots are trained before being unleashed on to the operational squadrons. What he meant was that we didn't want to be doing a classic training pick-up normally conducted at the underslung load park at Odiham.

'I think the best option is to have them rig the load in situ and attach a three-metre strop to the load.'

'Roger that.'

'Then you have a three-metre strop attached to the bottom of your aircraft. Then go into the LS in quick time. Drop off the stores, equipment, ammunition and take off really quickly.'

'Ok – how do we attach the EOD wagon?'

'You come up into the hover, taxi forward with the end of the strop at shoulder height. We then get the lads on the ground to attach the three-metre strop on the EOD to the three-metre strop on the bottom of the aircraft. You are out in thirty seconds, one minute max.'

'Hoofing, job's a good 'un.'

Satisfied that we had a plan that could work, I phoned next door to 3 Para and asked if Swanny could come in to the JHF(A) tent.

'Hello, I assume this is about our EOD.'

'Yes, mate, we have a cunning plan.'

'Is it more cunning than a fox who is professor of cunning at Oxford University?'

'Yes, mate, it is.'

I explained the plan to Swanny. It was outside normal practice and not a standard OCF load park plan.

'What we need is the HHI' – Heli Handling Instructor, the bloke on the ground who organizes the load-handling party – 'at Sangin to connect the two strops together underneath the aircraft. We'll put a Cyalume on the bottom of the strop attached to the aircraft so the HHI can see it and it doesn't twat him in the head.'

'Good point, we don't want a hook smashing into his face – that could really ruin his day.'

After squaring the plan with 3 Para, Daffers, Sam, Dan and I briefed with the AH and then headed off to the aircraft.

Sam and Dan arranged the back of the cab. It was essential that the hook was properly prepared. The cab also has three hooks that hang underneath it. These are used to attach various loads that otherwise would not fit in the cabin. Sometimes it is much easier and quicker to just stick kit in a net and attach it via a strop to one of the hooks. The underslinging capability of the CH-47 is second to none – in fact, a Chinook can lift another Chinook underneath it *in extremis*. You don't see that one on helicopter Top Trumps.

The fore and aft hooks are fixed on the outside of the aircraft. The centre hook is kept inboard in the belly of the frame. It's reached on the inside by lifting a large trap door. The outside door is opened by a ratchet mechanical system, which pivots the door forward, enabling the hydraulic hook to drop down. The cab was then loaded with the resupply by the guys and the MAOTs (Mobile Air Operations Team). Sixteen troops with all their kit boarded the aircraft and we were good to go.

I jumped in the right-hand seat. Daffers was in the left. During the start sequence we did our hook checks.

Dan said to me: 'Select centre, hook master live.'

Daffers reached up above his head, selected the centre hook on the four-position switch, which is the switching panel that controls all hooks forward, centre and aft for independent or simultaneous release. He flicked the hook master switch from off to live, and the hook was now powered.

'Centre selected, hook master live,' confirmed Daffers.

'Right-hand seat, release, release.'

I raised my right thumb to the top of the cyclic and pressed the load release button.

'Releasing.'

The blue centre hook light on the CAP illuminated, confirming the hook was open.

'Hook open,' said Dan.

'We have the light,' confirmed Daffers.

'Recycle,' said Dan.

'Recycling.' Daffers reset the hook master from live to off and back to live.

'Left-hand seat, release, release.'

Daffers now pressed the button on his cyclic and the CAP illumination process was repeated.

'Recycle,' said Dan.

'Recycled,' said Daffers.

'Arming the WOG.'

The winch operator's grip looked like a Scalextric hand control and allowed the crewman to operate the hook from the back.

'Releasing,' said Dan.

'Hook open. Resetting.' Dan then manually reset the hook.

'Checks complete, hook master off,' said Dan.

Daffers switched the hook master off and then Dan restowed the hook.

We lifted, and followed by the AH we routed to Sangin.

Tension increased as we let down to low level. This had to be slick otherwise we would be sitting in the hover over the EOD with our arses swinging in the breeze.

'Ok, I am Mark, this is Daffers, you are Sam and Dan.'

We braced ourselves as we came over the top and saw the now familiar sight of Sangin. I eased back on the cyclic, reduced the power and set up for the dust landing into Sangin.

'One hundred feet, thirty knots, in the gate,' said Daffers.

Dust landings were becoming second nature. Our focus this time was the pick-up.

'Wheels on, ramp down,' said Sam on the number one. The troops rolled off in a heartbeat. Sam chucked off the supplies.

The strop is an exceptionally strong, complex combination of thousands of fibres, encased in a flexible hard plastic shroud covered in a rugged durable nylon sheath. The hook was inboard. Dan attached the strop and then dropped the hook down. He fed the three-metre length through the gap until it was resting on the ground below the aircraft, with the Cyalume visible on the master link. This was the hook at the end of the strop that the HHI was going to attach to the strop on the EOD.

'Are we ready?' I asked.

'Good to go,' said Dan.

'Coming up.'

Through the green hue of my NVGs, I could see the EOD vehicle in my one o'clock next to the compound wall. We were going to be really exposed.

I pulled power and the aircraft lifted.

'Up gently . . . Up ten.'

Dan told me how high I needed to be to get the strop at the right height.

'Strop's coming clear.'

The end of the strop was off the ground.

'Up a further five . . . Cleared forward and right.'

As I looked down I could see the HHI, getting sandblasted by the full force of the rotor downwash. He was stood next to the EOD, the end of the other strop in his hand. His head was down as he battled to remain upright against the hurricane-force wind from the blades.

'Got the guy visual. Coming forward and right.'

We hover-taxied forward.

If they shoot us now we are fucked, I thought. We were out in the open, ripe for the picking. My heart was in my mouth. The only thing that was missing was a large neon sign flashing: *Shoot us now.*

'Visual the hooker . . . connecting now . . . forward and right to centre over the load.'

I positioned the aircraft directly over the EOD.

'Up gently . . . prepare to take the strain.'

Hearing his instruction, I raised the collective, increasing the power to the rotor blades to accommodate the additional weight of the vehicle.

Daffers monitored the engine temperature. The increased power caused the engine temperature to increase, so it was essential we stayed within limits. I sensed the aircraft working harder as we lifted the vehicles clear of the ground.

'Good load,' said Dan.

'Clear up and right.'

I increased power to make sure that we could clear the dangling vehicle over the Hesco Bastion wall surrounding the Sangin compound.

'Clear of the Hesco,' said Dan.

'Good job, lads, that was slick.' We were in and out in less than a minute.

I pedalled, turned right and accelerated across the wadi. Once clear of Sangin we started our climb to height.

The ton of EOD was hanging by a cradle of four strops at the hook end of the three-metre length. We ascended at 80 knots. Suddenly there was a judder through the airframe.

'What the hell was that?' I said.

'Shit, one of the cradle strops has snapped.' Dan looked down through the belly of the aircraft and saw the EOD spinning around, with the back end dropped down at an angle.

'Mate, it's spinning like a bastard.'

'Brilliant,' I thought. 'It's the middle of the night, I am on goggles and I have a one-ton spinning vehicle to get safely down.' On the bright side, I knew that we had time in the hover because nobody was going to be shooting at us.

We were ten minutes out from Bastion.

'Right, lads, how are we going to do this? We need to get this vehicle on the deck without trashing it.'

It would be pointless to have gone to all this trouble to then destroy the EOD on the pan at Bastion.

Sam said: 'We need to come into the hover, wait for the spinning to decrease, bring it down very gently.'

Dan piped up. 'But it's at a funny angle.'

'Ok, so softly, softly, catchee monkey.'

On our descent to Bastion we were cleared promptly into the load park. I brought the aircraft to a high hover around 125ft.

'Engines are good,' said Daffers.

'Are those lights backing your goggles down?' The lights of Bastion were having a detrimental affect on the NVGs. I was working like a one-legged man at a kicking contest.

'Yes, mate.'

Dan's head was down through the centre hatch, watching the load. 'Down twenty.'

He judged the height of the vehicle, above the ground, at

night, with two 2kg toilet tubes stuck to his head, and called it bang on.

'Down ten, gently . . . Hold it there.'

I slowed the rate of descent.

'C'mon, c'mon, let's just get this thing on the ground.'

The vehicle spun round. The timing was critical.

'Down five . . . nearly touching.'

The vehicle touched but it was still spinning.

'Still spinning, up, up, up.' I pulled power and lifted the vehicle up off the ground. As the vehicle gently touched the HALS it massively reduced its revolutions.

'Vehicle spinning less. Down five. Load touching, down gently. All four wheels are on. Down five and right.'

The hookers were cleared in to detach the load.

'Is it ok?' I asked.

'Yep, it's good to go.'

'Good job, gents.'

We repositioned to the spots to land on and shut down the aircraft.

'Are we ever going to get a job that goes completely to plan?' I said to Daffers.

'No worries, mate. You'll be back at KAF tomorrow,' he said.

Chapter 26

I was ready for a change of scene. It would break the det up, I'd get some washing done and I'd have some Internet time. But KAF had its drawbacks. We were living up by the Sea of Tranquillity, i.e. 'The Shit Pit'. There were signs everywhere saying: 'BIOHAZARD. DO NOT SWIM'.

'As if. Really? No swimming? Shame.'

The waft of the vast poo pond, where the whole of Kandahar Airbase's excrement ended up, pervaded my nostrils. We were constantly smacked in the face by the odour of approximately 3000 people's shit. I knew for sure that I was back at KAF. When the wind blew across it was so bad that you wanted to go back to Bastion just to get away from the smell.

The frustrations of life back at KAF took about an hour to really piss us off. It was frustrating because of the stark contrast between the two locations. We had been working like pigs in hostile territory and had come back to square away some basic admin and get some cash out. It should be a relatively painless and easy task. But Dan and I had a nightmare that morning trying to get our hands on some dollars. We had to have a little rant to Daffers back at the grots – the accommodation block.

'Get this,' said Dan. 'The cashier is only open the first day after epiphany after the third month of the fourth year according to our Lord St John. In reality it's open from 10.00 a.m. to 12.00 p.m. each day and all he has to do is sit in his office for two hours and serve his customers. I needed to get some cash out this morning, so I took a wagon and drove to the cashier. I got out and walked up to the window and said: "Right, yeah, hi. I'd like to cash this cheque, please."

"'The cashier's not here,'" said the tool at the window.

"'Where is he?'" I asked.

"'Oh, he's doing his phys,'" said the tool.

"'But it's during your opening times. Can't you do it?'" I asked.

"'Nah, mate, it's gotta be the cashier.'"

"'So you are telling me that the cashier isn't here, during his own opening times, and he's doing his phys when he could be doing his phys outside his own opening times?'" I said.

"'Yeah,'" said the tool.

"'Ok,'" I said and walked off fuming. Can you fucking believe it!' ranted Dan.

This was usual for KAF. Bastion was more operationally focused, whereas at KAF everyone was in their own little bubble. They were inside the perimeter wire and generally did what they wanted. There were clothes and electrical shops, a cinema, coffee shops, a Pizza Hut and volleyball pitches. There were folk who could easily spend four months in theatre, cocooned from the battles raging outside the camp, lulled into a false sense of security in the military version of Butlins.

There was an increased mortar threat. The mortar alarm would go off and everyone would hit the deck and don their body armour. But quite often those who had been in KAF too long were almost desensitized to it and just carried on oblivious, while the rest of the base lay on the floor with their hands covering their heads.

I had arrived back in time for Jonnie Porn's birthday. Jonnie Porn was a real character. He was named Mr Porn following a party that he attended dressed as the famous porn star Ron Jeremy. He had an eccentric look modelled on 1940s aircrew, a handlebar moustache with waxed tips pointing to the heavens. Sometimes I worried about him and his eccentricities. His clothing, language and general appearance were being swallowed up by a bygone era. He was, however, a thoroughly spiffing chap. We had both done the eight-week Qualified Helicopter Tactics Instructor (QHTI) course in 2003, we both had a love of tactics and mission planning, and we got on well.

We all joined him for some scran in the DFac. Life in theatre as usual was fast-action gag-fest-tastic. Bored aircrew were not dissimilar to bored children. Having been Forward since we arrived I was a bit out of the loop about what prank was currently in vogue.

I collected my dinner and headed to the table to join the lads. Mr Porn had waxed his moustache especially for the occasion. His best mate and a real good mucker of mine was Stu Hague. Classic good looks but bijou, he was like a little pocket-size version of aircrew. Jonnie and Stu were sat together with a few other lads from the Flight. Stu had presented him with a number of 'fork handles' in his pudding, in place of birthday candles.

I put my tray down. 'Balls. I need to get a drink.'

This was my first error. Never leave a tray of food unattended.

Stu wasted no time. The minute my back was turned, he picked up my little packet of cutlery and snapped it in half and then placed it back on the tray so I wouldn't notice it had moved.

I came back with my drink, sat down and opened up my cutlery to discover that it was now a handful of shards.

'What the fuck?'

The table burst out laughing.

'You guys are fucking hilarious!' I said. 'I can't be arsed to get another set.' The cutlery was stored at the other end of the Dfac from where we sat. It was a pain in the arse to have to get a fresh pack – that was the gag. I tried to eat my dinner with my half-sized cutlery but gave up in the end and eventually relented and fetched myself a new set. It was living among such comedy genius that helped the long det evenings fly by.

Daffers, Stu, Jonnie and I walked out of dinner together. I had snagged one of the JHF(A), tight-bastard budget, battered, old shitty Land Rovers which we used for all our crews and all of our kit. We were allocated two to look after the whole Flight. The Strike Command guys in their Harriers had about six air-conditioned Land Cruisers. Bastards.

Originally, there were three knackered Land Rovers in total. One was the designated duty aviator's vehicle, which he had to use to get to the HQ and back. He was constantly on call and carried a radio permanently. The other two long-wheel-based hard-topped Land Rovers were all we had to move all the equipment and crews to the aircraft and back again.

They were on their last legs and there weren't enough of them. All of the Landies were brought out by 18 Squadron during Prelim Ops. The Land Rovers then got pulled into the JHF(A) MT fleet. The Army Air Corps then turned up with nothing but a couple of bottle tops. They started bleating and dripping about how they didn't have any vehicles. Once they were centrally pooled the powers that be looked at the pool and said: 'Ooh, look, the Chinooks have got four. They don't need four. Give the Apaches one of those.' And so the haemorrhage of vehicles began.

Outside the engineers' and stores tents were two Humvees. These had been sat there for five weeks and hadn't moved. One of them was chocked with a block of wood. They were the truck version with cab backs and bench seats.

They were so big we would get all our kit and crew in them easily. It was annoying seeing them so neglected and under-utilized when we were driving around in pieces of shit held together by the dirt and not much else. 'I wonder if they work?' said Daffers to Jonnie Porn as we drove past them. It was like they were looking at us and winking.

'Hello, gents,' they were saying. 'Why don't you come over and use us?' We couldn't resist the temptation. What with possession being nine tenths of the law, we thought we'd have these.

I stopped the Landie. Stu and Jonnie jumped out. They started sniffing around them. Stu leant nonchalantly against the door.

The Humvees are built for the American military and they have to be able to be used by the lowest common denominator in the American military. That can be pretty low. They didn't have a lock, they didn't have a key, they were automatic and had a push-button start.

Jonnie Porn tried the door of one of them. The door opened because they didn't lock.

'Weehay, we are in, lads,' he said.

Stu tried the door of the other and it opened as well.

They tried the switches and twiddled the knobs.

They both looked perplexed.

'It won't start.'

'This one won't either.'

'Do you think it's the battery?'

Jonnie popped the hood and had a look around inside. 'The engine looks ok. It's in reasonable order. It has got oil.'

He went back inside and started fiddling again with the switches. 'That's odd, I can get the lights to work. It's not the battery.'

We could get the lights to work but we couldn't get the

engine to turn over. There was definitely electricity, but not going to the starter motor, so it was just a matter of eliminating the problems step by step.

Jonnie had a rummage around the dashboard and discovered a switch underneath the steering wheel. It was on a block screwed into the dashboard, and on the other side of it were the wires to the starter switch. In order to immobilize the vehicle the Americans had pulled out the wire connected to the starter switch. By reconnecting the wire to the block Jonnie was able to flick the switch and the HV's 5.3-litre engine burst into life with a thunderous roar.

'Fantastic, it works, it works!' yelped Stu with glee.

It was a very simple gearbox. It had park, drive, reverse and a great big lever for the handbrake. Looking cautiously around to make sure we didn't get rumbled, Jonnie climbed in one and Stu Hague in the other.

'Let's get them to the accommodation block,' said Jonnie.

Easier said than done. The trucks were armoured with 1½-inch-thick glass throughout, the windows couldn't be opened, and as the HVs had been left to go to rack and ruin the wing mirrors had fallen off. Driving them was a nightmare.

'Look at Stu,' I said to Daffers. 'He's such a shortarse, he can barely see over the steering wheel.'

'Mate, do you need a booster seat?'

'Ha fucking ha.'

Daffers and I followed them in the Landie.

Every time we arrived at a junction, Jonnie Porn had to open the door, lean out and have a good look around to make sure that the road was clear to drive onwards.

They were both shitting themselves. We could be nailed at any point by the military police. We had no legitimate paperwork or insurance for them and they were owned by the Americans.

When we all arrived at the accommodation we discovered that one of the vehicles didn't have a parking brake. When the engines were running the hydraulics worked, and the brakes were hydraulic so they worked. However, when the engine stopped running and the hydraulics stopped working, the brakes stopped working and the Humvee just rolled away. In the end we had to get a huge lump of wood to chock it with.

Stu and Jonnie ran excitedly into the grots and proclaimed: 'Chaps we have sorted the transport problem. Come and have a look at this,' gleefully rubbing their hands together.

The lads bundled outside to see the battered old Humvees parked up in all their dust-coated, junk-filled glory.

'Where did you get these?' asked Dan.

'Don't ask,' Jonnie Porn replied.

'Let's just say we found them,' said Stu Hague.

'Mate, how did you drive that here without a booster seat?' said Dan.

'Very original!' said a straight-faced Stu Hague.

It was just what we needed to lift our spirits and get rid of the boredom of KAF. We all chipped in with emptying them of all the rubbish.

'It's likely they are left over from the previous taskforce,' said Daryl.

'I don't know, guys,' said OC 'B'.

'C'mon, Boss, look at the number of vehicles out here, who is going to notice?' said Jonnie.

'I might even try and cobble together some UK MT paper-work to make them look semi-official then we should be ok.'

'Fucking hell, have you seen how much ammunition is in the back of this?' I said as we cleared rounds and rounds of bullets out of the back.

Out of the two of them we made one the golden vehicle.

Between the two we had one good set of wing mirrors and one good set of lights. We stripped down the one with the knackered brakes and added all the good stuff on to the other one.

Because it had a huge engine that guzzled fuel, we had to come up with a cunning plan. It wasn't as if we could just pop to the local garage and fuel it up with our credit cards. The fuel gauges on both were empty.

'I have a cunning plan,' said Jonnie.

'More cunning than a fox who is professor of cunning at Oxford University?' I said. (If it ain't broke, don't fix it. I never claimed to be an original!)

'It is indeed. We'll go to the fuel pits in the shagged-out old Land Rovers with a whole bunch of jerry cans, fill them up with fuel and then put it down as spare Land Rover fuel.'

'Brilliant!'

Jonnie Porn and I jumped in the Land Rovers ladened with jerry cans and headed for the fuel pit. The fueller was an older black guy, who used to be in the American Navy.

'I was a fueller onboard USS *Harry Truman*. I was bringing home $25,000 per year. I was approached by these guys and they offered me the same job, six times the salary. I handed my notice in to the Navy. The company then shipped me out to Kandahar for three years.'

'You are here for three years, mate?' I asked him. 'That's pretty shit.'

'I am here for three years, sure enough, and I get some holidays but the difference is when I get back home I won't have a mortgage and I won't have to work again! So who's the idiot here?' He laughed. The dude had a point.

He was so busy talking that he didn't care about the jerry cans. Fuel problems solved. We were laughing.

We were on a roll, it was a mission.

'Now, how do we get one of these beauties up to Bastion?'

We were even more strapped for decent vehicles up there.

'It's going to stand out like a pair of racing dogs bollocks in Bastion. I haven't seen too many Humvees up there,' I said.

We were more likely to get rumbled. We couldn't stop. The world was our oyster. How we could we undersling one on a cab without anyone noticing? The plot thickened.

'We'll undersling one. We did when I was on *Ark Royal* in 2003 as part of the ops going into Iraq. It was a doddle. We'll undersling it and we'll take it up to Bastion,' said Jonnie.

'Then we'll have one to drive around Bastion as well and we'll keep one here. Brilliant.'

The cabs going down to Bastion were full. We couldn't get the Hummer down there in the short term and so made the most of our new wheels more locally.

We tried to be reasonably subtle driving around. KAF was full of Americans. There were loads of Humvees driving around. We just blended in and nobody gave us a second look.

The thickness of the windows turned out to be a real advantage, that and the long sides. Driving along and sitting back meant nobody could really tell whether it was a Brit or American driver. It was so busy, with Humvees all over the place. What was another vehicle among hundreds?

As it was a four-wheel drive, it was awesome. It glided over the stupid speed bumps that were everywhere and it was air-conditioned. It was so big that we could get everyone into it much more easily than the battered old piece of shit Land Rovers.

We would go to lunch and dinner in it and all the REMFs would be queuing up complaining and muttering to themselves about our new wheels. We were riding the high life. It was amazing how much our spirits were lifted.

Daffers and OC 'B' were on the programme to do some tasking.

'Will you run us up to the light line?' asked Daffers.

'No worries, mate. Any opportunity to drive the beast.'

I jumped in the driver's seat, Daffers was next to me and his crew climbed in the back.

'Let's get the weapons and then go up to the line,' said Daffers.

I drove up to the armoury and parked up.

Daffers and I were chewing the fat in the front, while the guys loaded up the Humvee with rifles and ammunition.

A very large, black US sergeant came out of his office, looked at the Humvee, and walked up to it and knocked on the window.

'Excuse me, sir, where did you get the Humvee from?' said the portly African-American gentleman.

'Sergeant Kawalski said that we could use it.' I was winging it.

'Who is Sergeant Kawalski?'

My palms started to sweat. Shit, we are going to get rumbled.

'I don't really know him, tall fair-haired guy. Said it wouldn't be a problem. We just needed to run some kit to the cabs.'

'We have been looking for it for a few days now. We were doing an inventory because we are handing over to the next platoon,' he said.

'Really, right, well, we are just going to drop our kit off then you can have it straight back. All right, mate?' I said.

'Yeah, ok, sir,' he said.

'Uh oh,' I said. We drove back to the line. OC 'B' was still at the aircraft.

'The Americans want their vehicles back,' I told him.

'Really, but they have been sat there for ever!'

'I know but they are doing a kit audit.'

'I want to keep them,' Daffers chipped in. 'They wouldn't have noticed if it wasn't for the audit.'

'I don't know,' said OC 'B'. 'I think we should really give them back.'

And we all knew he was right.

We used it for the next half-hour or so and then with a sad heart handed both of them back. Although we were gutted to return them we were chuffed as punch we had cruised around in one for a week or so.

To add insult to injury, after we had returned the Humvees we found out that the Americans were dishing out a reward of $4000 for their return. We were the losers all round. Not only had we handed them back and not claimed the reward but we were back to driving around KAF in our two shitty Land Rovers. At KAF that sort of thing mattered. But I was moving Forward again.

Chapter 27

I walked into the CP at Bastion. McGinty looked up.

'Back again so soon? It feels like you've never been away.'

'Longest ten days of my life, being at the Rear. I am glad to be up here in the heart of the action.'

That was not strictly true. It had been good to be at KAF. I had made the most of the gym and stealing the Humvees had been a blast.

'We "borrowed" two Hummers from our American cousins. It was hoofing. If only we could replace all of those shagged-out Landies with Humvees we'd be laughing.'

'I wish,' said McGinty. 'Those Landies have got years in them yet. They'll still be here on your next tour, mate. When are you back out again?'

'March 07 is my next holiday at Camp Bastion. It works out as two months, every six months.'

'Are you back to do the RIP?'

'Yep, Jonnie Porn and OC 'B' are leading it.'

I had come Forward to take part in the biggest deliberate op to be flown to date by the CH-47 during our time in Afghanistan. We were taking in India Company 42 Commando Royal Marines, supported by 'B' Company of 3 Para and extracting 'C' Company 3 Para out of Sangin. It marked the beginning of the end of 3 Para's six-month arduous campaign in theatre. A *roulement* – a Deployment Rotation – between the two units was taking place across all the DCs throughout the province, in various guises.

Instead of taking everybody out in one go, leaving the

place empty and then putting the new troops in, we did what was called a Relief in Place. The plan was to put India Company in first, give them time to do a handover, establish the lie of the land and the enemy hot spots, and then when they had sussed it all out extract 'C' and 'B' Company 3 Para.

This was the first time an RIP of this size had taken place in a combat theatre. It was the first large-scale operation for us Brits in Afghanistan during such a high threat level. This was the beauty of being in the Chinook world – it was a richly diverse environment which continually tested our flying skills to the max.

The plan was to have a ground offensive while we were there unloading troops. The Paras in the DC were going to push out into Sangin town and then move to the south. This was to make sure that any Taliban fighters were cleared out to secure a cordon of a couple of miles around the DC so that we had a freedom of manoeuvre with no threat from insurgents.

We were going to take 42 CDO RM in and put them to the south-west of the DC, out on the dried-up riverbank. This was scheduled for daybreak.

The Royal Marines were to patrol down to the south through some old Afghan emplacements left over from the Russian conflict and then move around to the east to meet up with the Paras who had pushed out that way. Once they had made the RV they were all going to fall back into the DC together and begin the handover. We were then going to extract the Paras from the DC the next day.

There were so many crews Forward that we were accommodated in the overspill tents. These were not as well equipped as the IRT tents, just eight cots and none of the other comforts of home. I had just dropped off my kit in one of them and wandered out for a smoke. I saw Daffers.

'Do you know what the crack is, mate?' I asked.

'No. Jonnie Porn is leading. He should be over in JHF(A) planning.'

'I am going to go over and get a handle on what's going on.'

I wandered back over to JHF(A). Jonnie Porn had his head down, surveying the maps at the bird table.

'Mr Porn, how's it hanging?'

'Mate, this is a fucking deconfliction nightmare.'

Jonnie Porn was the squadron QHTI. The course we had done together aimed to provide frontline helicopter squadrons with instructors who were experts in helicopter threats and defensive aids. We were schooled in deconfliction, and the planning and conduct of complex, multi-aircraft missions. This was a tactics instructor's wet dream.

'Why, what ya got?'

'We have three waves of five aircraft into Sangin. Phase one is at first light, phase two is around 10 a.m. and phase three is around 1300. Guns to the south in FOB Robinson, we have got the MOG on the western bank, we have Teletubbies crawling all over Wombat Wood.'

Wombat Wood was to the north of the Sangin DC on the northern banks of the river running east–west. The wood's trees and bushes provided cover and thermal protection for the Taliban, hiding them from the sights of the Apaches and UAVs (unmanned aerial vehicles, assets which send back video footage of enemy positions and targets). Anybody who was moving around in the woods was very difficult to spot. The Taliban would use the woods to set up their mortar bases, from where they would fire on to the DC and any incoming aircraft. This was one of the places that we had been engaged from on the IRT shout on the 6th. If we had multiple helicopters on the ground in the DC and it came under heavy mortar attack from Wombat Wood, then the RIP would be a catastrophe of epic proportions.

'Mate, every time I have gone into Sangin, we have come in from the western side.'

'That's fine, but that's on your tod and we have got to get five in three times.'

'Are we going to the DC?'

'We are for phases two and three but phase one is dropping your brethren off on the western bank so that they can patrol down and push the Teletubbies south in a pincer movement.'

'You've spoken to the colonels?'

'Yessss,' he said sarcastically.

'What's their take on it?'

'They want all the pieces in place so if it kicks off their guys are protected. They have the guns to the south and the MOG to the west. Leaving us the problem of getting in but also leaving clear arcs of fire for the ground troops. The last thing we want to do is bong a gun target line. Then we'll look like a right bunch of muppets.'

The one thing that couldn't happen would be to leave a five-ship formation through a gun target line causing the guns to check fire in the middle of a potentially crucial fire mission.

'Where are you putting Royal into on the western bank?'

'Around here.' He pointed to a position on the map. 'Tootal is concerned about mines in that area.'

'Mate, we took a two-ship into this area here. We dropped Bill Neeley into there. It's further south but that could work because we know it's clear.'

'Thanks, that's good gen. Let me check. I'll run past OC 'B' and we can have a chat with the Colonel.'

Jonnie Porn had a task ahead of him. It was about trying to strike a balance between where they needed to be, where they could be and where we could get them in. In order to implement the RIP we had to weave and negotiate our way through a giant ammunition obstacle course. There were the Taliban, our own guys and then a grid of lines drawn on to a map, which either couldn't be crossed at all or only at certain heights.

There was Intelligence that the enemy was housed in emplacements left over from the Russian conflict. It was suspected that they had a heads-up that there was a changeover coming and were in place waiting for us as we were coming in.

An RIP exposes the helos and the DC more than normal inserts as the transition creates patches of weakness. There are guys that are now beginning to wind down because they are knackered and focused on getting home, so they are liable to get shot. There are also guys that have never been there before who don't know what to expect and don't know yet what they have to do, which makes them vulnerable as well.

Jonnie Porn was having difficult conversations with the Fires Officer. This was a guy who was responsible for co-ordinating the artillery and mortars for the op. He wanted pre-emptive fire, and while this lent itself to the IRT shout in Musa Qala, it would cause a problem for Jonnie because it would be a combat indicator and let the Taliban know that something was about to happen.

Jonnie did a great job convincing them that a quieter approach would offer a greater chance of getting the helos in and out unscathed. His plan was to come in as stealthily as possible, under the cover of darkness, without a fanfare of artillery. This would be done in one wave of five with around fifty guys on each cab. This was phase one.

The day before, there were Battle Group Orders, to ensure that we were singing from the same hymn sheet. They were given by Tootal and took place outside the JHF(A) at Bastion under the shade of desert camouflage netting.

There wasn't enough space to accommodate all the parties in any of the command tents. We placed some benches in rows in front of a huge briefing board with a large map clipped to it. We used clear acetate sheets with different grid lines, artillery, mortar and helo movements each on their own clear sheet. As the briefing continued we could put the sheets one on top of another and gradually build up an overall picture of how the RIP would work.

The brief was given in the NATO standard sequence of orders. Tootal began. 'Welcome, gentlemen. These are orders for the Relief in Place of Charlie Company 3 Para by India Company 42 Commando Royal Marines in Sangin.'

Colonel Tootal then covered Ground, Situation, Enemy Forces, Situation Friendly Forces and the Scheme of Manoeuvre. During this briefing process a representative from each group that had a part to play stood up and summarized their game plan and what they were doing. This gave an overview of the whole picture.

Having come from a Commando background I am used to this sequence of orders. Having been excellently briefed, we were under no doubt about what was required. It was a big push and it had to be as well prepared as possible. We then broke off for the formation brief. OC 'B' took the floor and outlined who was crewing each cab.

'In ten seconds the time will be 1224 and twenty seconds. In five, three, two, one, hack!'

'Call signs, Footloose 24, 25, 26, 27, 28.'

Everyone replied in turn.

OC 'B' led the briefing because he was leading the overall push. Jonnie Porn was the deputy. It took place around the bird table. The J2 gave a brief update on the Intelligence and the contacts in Sangin.

The brief given by OC 'B' followed the standard SOP18 brief, a formula for crew which ensured that all the relevant information was passed on to the crews flying the sortie. He was leading the three-ship and Jonnie Porn was leading the two-ship. OC 'B' went through the precise time that our wheels were going to touch down, the exact heading that we were landing on and the way that each aircraft was going to face. We were coming in from the north-west so we landed facing south and then were to fly out to the west.

The Apaches' role was as ever to protect us. They were to be our eyes and ears for all phases, covering Wombat Wood and other known enemy locations, and neutralizing any Taliban that were brave enough to take a pop at us.

'Hammond, you are back-up for tonight with Kluthie. You will be Dash 3 on both waves tomorrow with CO JHF(A),' he said. This meant that we would be the third aircraft, with the lead formation, on phases two and three, flying with the CO. He was in charge of all the helicopters assigned to RC (South), covering the Chinooks and the Apaches. It was a rotational billet that up until then had been filled by an Army Air Corp regimental Commanding Officer or a Chinook squadron officer commanding. In this case it was an RAF Wing Commander, another highly competent operator.

OC 'B' told the rest of the guys their aircraft and spot positions. He ran through a timeline for the next morning.

'Check in at 0410, Fox Mike, Victor, Uniform,' ensuring that everyone knew which radios were being checked first, second and third.

The whole thing was kept deliberately low key. Little about the brief betrayed the fact that we were about to fly five aircraft, in three waves, into hostile enemy territory to do one of the biggest gigs of the entire detachment to date. But that went without saying. We all knew this was the big one.

We broke off and I had a quick chat with my crew. Kluthie was a young lad who had not been in that long but he had a few tours of Iraq under his belt and he was a good pair of hands. He was my co-pilot. Down the back we had three crewmen manning each of the guns. Mr Mack was on the rear gun, Spence on the left and Sam on the right. Lessons had been learnt from our night at Musa Qala and nobody was taking any chances. Each cab in the sortie had three crewmen aboard.

'Don't think that we'll be doing that much tonight as we are only the back-up,' I told them. 'Kluthie, sorry, mate, but you are being kicked out for CO JHF(A). RHIP, my friend' – rank has its privileges.

I took my crew into a briefing with CO JHF(A) to run through the plan for the next day.

'Right, Boss, I take it you'll be in the right and I'll be in the left.' I knew that the CO would do most of the flying. Because of his position he needed to grab every opportunity when it presented itself.

'We'll be Dash 3 for phases two and three. Everyone should be happy with the routes and the outline plan. It just leaves me to cover emergencies and actions on. Emergencies will be dealt with using the WADFIR principle – this is Warn Crew, Actions (immediate), Diagnosis, Flight Reference Card's, Intentions Radio. These are the actions that we will take in the event of an aircraft emergency.

'Is everyone happy with the tracer and contact call?'

They all nodded. Tracer meant that bullets are being seen but are not affecting our aircraft. Contact meant that it was effective fire that would affect the aircraft. Even though we had come under fire many times this det, to date none of us had called 'contact'.

We covered the ABC of evasion. A equals 'action' – 'break

right', 'break left', 'continue'. B means 'because' – either 'contact left', 'contact right', 'RPG' or 'small arms'. And C equals 'clock code' – one o'clock, etc. This principle had been used every time we had been shot at and had saved our lives at both Sangin and Musa Qala.

The brief took place the day before because we had five crews that were all required to be in crew rest prior to the beginning of the RIP. We had to anticipate flying for a whole day past the initial insert. The time on target for the troops on phase one was 0501. In order to achieve this wave one had to take off at 0430. They needed to start the aircraft at 0400, and get up at 0230 in order to get the briefing done. In order to get eight hours' worth of crew rest they needed be in bed by 1830 that night. That's why we did the orders in the afternoon – to guarantee those eight hours' rest and getting up at 0230 to start the RIP.

Loading a five-ship formation required a great deal of logistical co-ordination of the troops. This stemmed from the orders process. Each packet of troops (chalk) needed to be on the right aircraft. The troops would arrive at 0230 to be led out to their respective aircraft spot by their chalk commander. Getting on the right cab was essential. If the chalk was on the wrong aircraft it would screw the whole plan up.

The troops were heavily laden with ammunition and weapons as well as their Bergens. Some were having a quick smoke before the aircraft were fired up. The aircrew were completing the updates briefs before going down to the aircraft. They filed out of JHF(A). Some walked over the rough desert, some piled into the Landies. The pilots climbed into the cockpits, the crewmen into the rear of the cabs. With a few seconds between each start sequence the five CH-47s powered into life. The thunderous roar of five engines simultaneously combusting deafened the troops waiting to board.

The lights of Camp Bastion shimmered in the background. The troops piled on and shortly afterwards the five ramps lifted.

Co-ordinating a take-off in a five-ship formation was bloody hard work. I wasn't flying phase one but I knew how it would be for them. All five cabs checked in on the correct radios and in sequence. At the designated time each Chinook lifted and headed north. In order to achieve the level of surprise that was required, the Chinooks had to come in first and the escorting Apaches would hang back.

Flying at night in a multi-aircraft formation is arduous, especially for the tail-end Charlie. It is a fine art to judge distance and stay in formation. Too close and not only will the crew start getting nervous, but there will be a real risk of a mid-air collision. Too far away and there's a risk of losing sight of the next aircraft in the formation.

They all had formation IR lights on so they could see each other on the NVGs, but they could also see the glowing embers of the burners at the rear of the aircraft, which ran at 700°C glowing very brightly. Even in the pitch-blackness of the Afghanistan night it was hard to distinguish the glow of the exhausts by the naked eye from the ground. There might be a glimmer of traces as the heat stuck to the sky but nothing to blow their cover thousands of feet above the earth.

As the formation was flying along it was surrounded by the amplified sharpness of the stars in the night sky. Even though they were transiting towards an intense operation, it was impossible to escape the wonder of flight. Looking through gogs, enveloped as far as the horizon by stars that aren't visible to the naked eye and coated by the inky black sky is one of the few feelings you can have of true freedom, where for a moment you are completely divorced from reality.

OC 'B' and his co-pilot, Daffers, were co-ordinating the

run in. The formation had reached the release point but needed to hold at height in order to achieve the time on target of 0501 before letting down into low level. The five aircraft descended. The first three leading were separated by four minutes from the following pair. Skilled formation pilots trust the aircraft ahead of them. As the lead started its descent, the crew's prime concern was to ensure that the timings were accurate in order to make the time on target. The second aircraft was focused on maintaining visual contact with the lead. The third aircraft was focused on maintaining visual contact with the second. This dance was performed to make sure that, once low level was reached, all aircraft were the correct distance apart to be able to land simultaneously on the LS. In the back of everyone's mind was that this had to be done swiftly so as not to fuck up the next element's landing times.

It was a textbook demonstration of a three-ship simultaneous dust landing – the third aircraft touched down seconds before the second aircraft, avoiding its dust cloud, the second touched down momentarily before the lead aircraft. The landing was synchronized and executed to perfection. Spectacular. The Royal Marines ran off into the pitch blackness of the Sangin night. Thirty seconds later the ramps came up on all three aircraft, which lifted and departed for Bastion. Two minutes later another two-ship formation executed an equally perfect insert and headed back to Bastion. Phase one was completed.

Chapter 28

I was flying phases two and three. I met up in the CP with CO JHF(A).

'Are you ready for this, Boss?'

'I am looking forward to it. It's going to be great to be flying.'

This was one of the few chances for the boss to get away from the chains of his desk and get back in the cockpit.

Kitted up, the boss and I walked down to the cab together. All five cabs were lined up on the pan, ready to go. It was gin-clear, the day was warming up to the Afghan ambient autumnal temperature of a balmy 25°C.

'So, Boss, when was the last time you did a desert underslung load?'

'It's been a while.'

I didn't want to come across as patronizing and he was an experienced aviator and had probably done more underslung dust desert loads then I had. However, he was bound to be a bit rusty. Not that I had major concerns but I knew I had to keep my wits about me. The plan called for us to take in a 4-tonne underslung load, land it on, go forward and then plane 35 Para and a quad bike back to Bastion. The loads were all prepped in the load park.

In the brief it had been decreed that the formation would lift in sequence, and individually move round to the load park, pick up the load and move to the north, prior to joining up in formation. The Apaches would again act as escort, flying tight with the formation as we routed into the DC.

The boss and I climbed in the cab. I was in the left, he was in the right. Minutes later we were up at flight idle and were ready to lift.

OC 'B', as the formation lead, spoke on the radio: 'Tower, Footloose 24, request reposition to the load park.'

'Footloose 24, Tower, reposition as requested.'

We watched as his cab lifted, routed in front of us and flew round to the load park to pick up his load.

Moments later: 'Tower, Footloose 25, request reposition to load park in turn.'

'Footloose 25, Tower, reposition as requested.'

The second cab lifted, routed in front of us and headed to the load park. It was now our turn. I transmitted on the radio: 'Tower, Footloose 26, reposition load park to follow Foot-loose 24 and 25.'

'Footloose 26, cleared as requested.'

I sensed the controller in the Tower was getting bored. We moved over to the load park to pick up our load. Any misgivings I may have had about the CO's capability soon evaporated, when he performed a textbook load lift. We joined up with the rest of the formation and routed north to Sangin. En route I asked Mr Mack, who was on the number one: 'Have you got sight of Dash 4 and 5.'

'Yes, mate, they are moving round to the load park now.'

Excellent, the separation was bang on between them and us, I thought.

We routed up to the release point. 'At the release point now, OC 'B' should be dropping down to low level shortly. There he goes. Follow him down,' I said.

'Descending,' said the boss.

'Going through 1000 feet . . . passing 500 . . . 200.'

'Pulling in power.'

'Two good engines.'

We were now in low level and headed into the compound of the DC. We were lined up in Echelon Port and began the descent into the compound. Sangin's compound was large enough to easily accommodate a single Chinook but we were now a three-ship, each with 4 tonnes underslung, planning to land at exactly the same time. Our formation distance had to be maintained throughout the entire landing.

This should be interesting, I thought. *Where are those Teletubbies?* It was making me more nervous that we hadn't experienced any contacts yet. A three-ship was a fairly obvious target which had limited room for manoeuvre.

'If they took a pop at us now left's our only option. ' I was scanning the horizon for the bad boys.

We came down into the gate.

'One hundred feet, thirty knots, in the gate.'

The CO had his eyes on Dash 2 and the lead. It was crucial that we got our load on the ground before we got caught up in Dash 2 and lead's dust cloud.

The RADALT audio warning sounded in our headphones: 'Cancelling.'

Mr Mack picked up the load descent patter. 'Twenty feet below the load . . . ten feet steady, down ten for the load . . .'

We descended.

'Load's on the deck . . . load's gone.'

Mr Mack released the load and the aircraft instantly felt lighter.

'Going forward and down. Landing on.'

We were then engulfed in a second dust cloud as we transited forward. As the dust cleared, past the CO I could see Dash 2 and the lead on our right-hand side.

Instantly thirty paras swarmed on to the back of the cab, eager to get home and put Sangin behind them.

'Quad's coming,' said Mr Mack.

A para on a four-wheeled motorbike drove on to the back of the aircraft. It never ceased to amaze me how slick the paras' helicopter and planning drills were. As a Bootneck, admitting that the paras were good at anything was like choking down a bowl of Marmite. Within forty-five seconds we were ready to lift. The three aircraft were all facing south, line abreast. I watched as lead and Dash 2 lifted in turn and routed out to the west. The aircrafts' departure was elegantly choreographed.

'Lifting,' said the boss.

We followed the other two cabs, routed west and climbed to height. We were relieved that phase two had been completed so smoothly. One more phase and we would be home and dry.

We landed on, offloaded the troops and shut down the aircraft. As the boss and I walked off the ramp, the paras that we had just brought back were still around the rear of the cab, waiting for transport. I was shocked to see what rag order they were in. Their combats were rotting off their bodies. They were pale and emaciated, with long straggly beards. They had clearly had a rough time of it and they looked knackered. I had immense respect for these guys. What they had been through and what we had done in the last four weeks paled into insignificance compared to their six-month ordeal.

The boss headed to the 3 Para galley for some scran. Some of 3 Para were there ahead of us. It was a testament to their levels of exhaustion to see that some of them had fallen asleep in their lunch. It had been a toss-up between some good food and their racks. They chose food. At the end of the day paras are still paras.

Chapter 29

We were feeling bullish at our success so far as we sped along at height with another 4 tonnes underslung and a cab full of Bootnecks, laden with massive Bergens, webbing and ammo. The troops were chomping at the bit, like racehorses in the gates, ready to get in the game. It was the final part of the Sangin RIP.

The day was cooling down. Throughout the day India Company had pushed out with 'B' Company and had been engaged in sporadic firefights, pushing the Taliban away from the DC. As they folded back in preparation for the pick-up, the Taliban had closed in with them.

'We are at the release point,' the radio crackled.

'Boss, the IRT has been launched to Sangin. We may be holding,' I said.

We watched as lead and Dash 2 went into a right-hand orbit.

'Yep, we are definitely holding.'

Shit, I thought, *this is bad news*. It spelt trouble, first of all one of the troops was injured, which was never good. Secondly, we had been into Sangin twice now, with five helicopters. We were going to send another helicopter in, followed by a further five helicopters. The success of the whole operation had relied heavily on a quick in-and-out. Unfortunately, we were now a formation flying at height clear enough for the effective Taliban dicking network to alert their friends at Sangin that we were on our way. It had the potential to get very messy.

'I can see the IRT coming out from Bastion.'

'Who's the Captain?' asked the boss.

'Stu Hague,' I replied.

The IRT cab was racing towards Sangin.

Why don't we just send them a fucking postcard? If this isn't a combat indicator to the Taliban, I don't know what is, I thought.

This put us all on edge. We just wanted to get on with it but we had to wait for the IRT. The whole operation needed to stop to ensure that the IRT got the casualties and got them back to Bastion. We all understood this.

Jonnie Porn's section, Dash 4 and Dash 5, had inevitably caught up and now joined the holding pattern.

'You have gotta see this, Dash 4 is shedding its load.'

Jonnie Porn's load was made up of rations and water in a net. Although the load was heavy it was loosely packed and the ration boxes were beginning to break open, showering good old British Army rations all over the Afghan desert.

'That's one way of feeding the enemy,' said Mr Mack.

It caused some temporary light relief. We were holding for ten minutes as the IRT flew in to pick up three casualties.

'I think that's IRT coming out now,' I said.

'We should be pushing in shortly.'

No sooner had the words left my lips than we saw the lead aircraft begin its descent. We were left with no other choice but to adopt the same plan from phase two – three aircraft, line abreast, with a 4-tonne load hanging underneath, followed by a two-minute separation for Dash 4 and 5. We dropped down to low level all into Echelon Port, three heavily ladened aircraft, trapped in their own airspace, slowly descending towards Sangin DC.

Daffers and OC 'B' were in the lead cab, with Tony on the number two. All of sudden everything kicked off. An RPG flew down the right-hand side of their cab.

Daffers called it. 'RPG right one o'clock.'

It exploded 200m behind the aircraft. We couldn't make many manoeuvres as we were trapped in the valley in formation on short finals to the DC. The shooters were so close that Tony couldn't depress the weapon far enough down to fire on them. The Taliban had fired an RPG 50m from the side of the aircraft. A second RPG was fired across the front of all three aircraft. I saw something that looked like an American football being thrown really quickly. It passed the windscreen and then exploded. There was nothing I could do. We had experienced so many RPGs this det, it was unreal. In a deadpan voice, I calmly said: 'Continue, airburst RPG, my eleven o'clock.'

Daffers called the Apaches on the radio: 'Wildman, contact, my right three o'clock, two Taliban fired two RPGs.'

The Apaches rolled in. They missed them on the first burst. As I looked down I could see a young fella and an old fella both carrying firearms. The young fella sprinted off while the old fella slipped up and fell over in his sandals. The young fella stopped, looked back and then carried on running. The Apache tracked the young fella with the gun. A burst from the 30mm cannon went off around him, he fell over, got up, ran about ten metres and realized that his number was up. He fell over again and lay still while the Apache hosed him down with bullets. The old fella got up and ran into the treeline, closely followed by the Apache, who hosed the trees with bullets. They had got them on the second. It was clear that the Apaches had well and truly wasted them.

We continued the descent, dropped the load, moved forward, landed on. Our hearts were racing. We wanted to be in and out of there quickly. We couldn't afford to waste any more time. The enemy was everywhere and closing on the

DC. The troops on the ground were engaged in heavy fighting. The ramp went down. The troops ran off. One Royal Marine on Dash 2 misjudged his exit off the ramp – weighed down with a heavy Bergen he slipped and snapped his leg. This poor Royal Marine, who was fully psyched up to get in the thick of it with his oppos, was now hauled back on the helicopter injured to be taken back to Bastion.

The paras began to quickly load on. There was a heavy ditch going through the centre of the compound. We were briefed to take a quad bike back. In his haste to get on to the aircraft, the young para driving the bike forgot about the ditch and drove the quad bike into it, causing it to stall. He frantically tried to restart the quad.

But we had been on the ground for too long and the cab couldn't wait for him. The para pulled his quad out of the ditch and managed to restart it. He drove it as fast as he could towards the ramp but there just wasn't enough time. We had to lift and did so before he could make it. Mr Mack was shouting for him to jump but it was too late. His train had left the station. This para was stuck at Sangin, among his new Royal Marine friends, until we could go back in and get him out on the next resupply.

Mr Mack felt bad for the poor guy but there was nothing he could do so he stood on the ramp and waved fondly at him as we ascended away from Sangin. He was the only para in a place which was now full of marines and it was all because he had stalled his quad. It became a running joke. 'When are we going back to get bike guy?' He was there for a week. When he finally got on a cab, Dan said smirking: 'Are you glad to be going home, mate?' The para met Dan's eyes. If looks could kill. Clearly he was not very amused at all.

We landed on Bastion and shut down. The operation had been a huge success. All the crews converged on the JHF(A) for a quick debrief. While we were waiting for OC 'B' to start the debrief the Apaches showed us the gun tape. Daffers took the piss out of them for missing on the first pass. The gun tape showed that they had missed them by 30m.

'I thought you guys were supposed to be good,' we laughed. 'You were fucking miles away.'

The debrief went well. We had all got in safely and there were no major issues. Both the insert and extract were seamless. Jonnie Porn had done an excellent job.

At the end of the debrief, Daffers, Dan and myself worked out that we had had a total of seventeen RPGs fired at us over the course of the det. We were trying to work out who was the magnet. My money always was and always will be on Daffers.

Chapter 30

I was stood with the other guys around the bird table in the CP. The 3 Para Ops Officer was explaining to us the mechanics for the planned withdrawal of the guys from Musa Qala.

A truce had been brokered by the tribal elders with the Taliban, to guarantee the safe withdrawal of 3 Para's troops from the DC. The plan was to take them out of the front gates in a convoy of jingly trucks. However, Musa Qala had been a turbulent, besieged outpost deep in the heart of Taliban territory.

During the early stages, when the footprint of the Afghanistan stabilization campaign was being laid, a platoon of 24 Pathfinders, a deep-reconnaissance force drawn from the paras, were the first British unit to engage with the Taliban in Helmand Province. They were providing back-up to the Afghan Police, under attack in the DC.

A month later, in what was supposed to be a two-day mission to hold the fort until a company from 3 Para battle group could get in, they were stuck still in Musa Qala. Their ordeal lasted more than seven weeks. They were totally surrounded and entrenched in fierce fighting, while we were looking for a way to get the paras in. For the next fifty-two days the Pathfinders fought for their lives under the relentless Taliban assault.

The Taliban had regained control of this part of northern Helmand Province. But President Karzai's government wanted to fly its national flag over the four district centres of Sangin, Musa Qala, Now Zad and Gereshk, rather than the white flag of the Taliban.

The local residents of Musa Qala were reluctant to be sucked into the violence. In front of the District Centre was the main bazaar and trading area of the town and they didn't want to see it damaged by heavy fighting.

This meant that at first the Pathfinders had freedom of movement and were able to set up vehicle checkpoints and mount night patrols. In the meantime, the engineers from 9 Para rebuilt the District Centre's defences and repaired its sanitation.

Int were getting reports that the Taliban were determined to retain control of the town because it formed part of the route into Sangin. In a bloody brutal display of aggression the Taliban forced a young woman, accused of assisting co-alition forces, to watch as they killed her young son. Then they publicly hanged her. They meant business and they wanted us to know it.

The Taliban often put on a show of strength by roaring up and down the town in 4x4 pickup trucks full of their fighters. Soon the Pathfinders' vehicle checkpoints and night patrols had to be withdrawn.

The DC was attacked daily with everything the Taliban had in their arsenal: rocket-propelled grenades, Chinese rock-ets, snipers, mortars, heavy-machine-gun fire, small arms and their favourite – noisy, forbidding recoil-less rifles.

The Pathfinders were shocked to find themselves under constant and deliberate attack. They were literally firing tens of thousands of rounds from every weapon they had. With a 24-hour guard, they sat in protected bunkers scanning the town with binoculars, being blasted by mortars with follow-up attacks mounted by many Taliban, and firing thousands of rounds back at them. The fighting in Musa Qala had become very intense.

Luckily, the Pathfinders were experienced soldiers but the

might of the Taliban assault was not lost on them. The Taliban refused to back down, even in the face of the Pathfinders' withering response, which included calls for support by means of air and artillery strikes, killing hundreds of Taliban. But the attacks continued. The Taliban had control of the roads, and by targeting the helicopters they gave us little chance of resupply. Ammunition, food and water needed to be rationed. The Pathfinders simply had to make every round count.

They believed fiercely in their chain of command and knew that relief was on its way. They were living on their wits, sleep-deprived and with little food. They were snatching catnaps under cardboard ammo boxes, clutching their weapons, in rotting clothes, and then engaging in heavy gunfights, while the Taliban continuously bombarded them with RPGs and mortars.

Luckily, the engineers had fixed the well inside the DC compound, which provided a limited supply of water, otherwise it would have been over quickly. A month into the siege, a relief column of Danish armoured reconnaissance troops backed up by British signals specialists left Camp Bastion to try to take over from the Pathfinders.

The column was attacked from three sides at once when it reached the outskirts of Musa Qala. The Taliban engaged it with heavy-machine-gun fire and rocket-propelled grenades, and blocked the road with oil drums and barrels.

A landmine destroyed a Danish Eagle armoured vehicle, wounding three troopers. The Danes opened fire with their own heavy machine guns and called in an American B-1B bomber to bomb the fuck out of the Taliban before moving back into the desert to regroup. Inside the compound the food and water supplies had run out, and the Pathfinders were now surviving on goats' milk. They were somewhat concerned by the Danish withdrawal. The Danes finally managed to push into the compound and were greeted by a group of filthy, flea-ridden, emaciated and bearded soldiers, all grinning like schoolboys on Christmas Day. It could have been a scene from *Lord of the Flies*.

The celebration was quickly interrupted by another Taliban sniper attack, shooting a young Sergeant from the column in the head. Luckily he survived the potentially fatal wound. The next morning the Sergeant was picked up by a British Army Air Corps Lynx helicopter, which swooped in and landed at great speed – an essential tactic to ensure that it didn't come under attack.

However, the enemy had retained control of the surrounding area. The Pathfinders attempted to fight their way out, supported by a troop of British Scimitar armoured reconnaissance vehicles. Inevitably, the Scimitars were ambushed and destroyed. Tragically three British soldiers were killed in that contact.

The ambush on the Scimitars showed us that only an overwhelmingly strong force would be able extract the besieged Pathfinders and reinforce the Danes.

When the extraction and reinforcement op began, the Taliban positions had already been battered by B-1 bombers, the night before. The British commander, Brigadier Ed Butler, ordered 300 paras to be inserted by helicopter, around half a mile outside the District Centre, to assault the Taliban. It was to be a classic infantry charge with bayonets fixed, behind a barrage of artillery and mortars.

When they got to their objective, the Taliban had fled, leaving the compounds empty. Only a few stayed to fight and there was some sporadic small-arms fire. Tragically a para private was shot dead. But the relief column had finally reached the beleaguered troops.

Two platoons from the Royal Irish Regiment, under the

command of Colonel Tootal and the 3 Para battle group, replaced the ecstatic Pathfinders, who left in Land Rovers smiling and waving. Shortly afterwards, the Irishmen found out why the Pathfinders had been so delighted to be leaving Musa Qala.

The Royal Irish had been lulled into a false sense of security by the town's deserted welcome, before the Taliban unleashed a barrage of attacks. By the third day two soldiers with gunshot wounds had to be airlifted out of the base.

Four weeks later the Danish squadron left and were replaced by another Royal Irish platoon. The Taliban continued their offence with a series of massed attacks, which were repelled with the back-up of US and RAF jets.

The siege continued, and in four weeks of fighting the Royal Irish had alone fired 25 per cent of all the 7.62mm machine-gun rounds used by British troops in Afghanistan throughout the whole of 2006. There were more casualties and unfortunately some more deaths.

On the 6th we had been called in to pick up the casualties and had the shit shot out of us. As a result we hadn't been able to send another cab back in after 6 September and the situation for the troops on the ground had got gradually more intense.

The elders were anxious to restore normality and so approached the Brits with an unusual offer. It was agreed that if we stopped fighting the Taliban in the town, then the elders would force them to leave. The elders had effectively negotiated a ceasefire.

From an aircraft perspective there were to be three Chinooks on standby, turning and burning on the ground in the desert out to the west of Musa Qala while the troops were driven out on jingly trucks. The Chinooks were to be loaded with reinforcements on the back for additional support.

All the contingencies we could think of before we lifted to reposition the aircraft to a deserted patch of nothingness were briefed. As usual, among the deadly serious business of discussing our options under fire, there were the odd comments about people's perceived bravery, or lack of it:

'Thank fuck Hammond's on IRT. If he was in the formation, it would all kick off. He is a nutter and will probably get us all killed,' came the banter from the back of the room.

Hoofing, I thought, *my reputation as a Bootneck is thriving.*

'Yeah,' the banter continued, 'as he's a Bootneck, surely he can't even read a map. We'd all end up in Mullah Omar's house somewhere near Kandahar!'

Ho ho, I thought. *There I was thinking it was because I was tough as old boots, but no, really it's because they think I can't spell.*

They crewed in as usual, gathered their well-armed passengers and sparked up the cabs, with a thunderous roar. The engines pumped and the rotors tumbled round at 225 revolutions per minute. Emotions were running high as we knew the extraction could turn into a massive firefight.

We were poised and ready to deliver the reinforcements straight into any contact, before removing casualties. We also knew that nothing might come of it and we would probably end up sitting in the desert for hours playing I Spy. There were only so many times you could say 'something beginning with S' before people wanted to start punching you.

As they approached the desert Laying Up Point (LUP) they set up for a three-ship desert landing. The third aircraft has to be on the ground first or it ended up eating the dust cloud of one and two ahead of it. This was challenging flying, especially for the number three cab pilot, who woud sit back in Echelon Port watching for the exact moment the lead cab started its approach. They landed on gently enough not to break the wheels off. This landing went off without incident and they sat there in noisy solitude wondering what was coming next.

For the guys down the back, the magic vibration of the Chinook had set in as soon as they took off. It was a strange, hypnotic pulsating that seemed to have the power to send all the passengers to sleep, no matter how hard the cabs were flown at low level. Passengers had been known to sleep through a contact with rounds flying around, and occasionally through, the aircraft and inevitably to wake up with a stiff neck and drool stain on their kit.

The crews played stupid games and engaged in remorseless banter. Banter was our way of expressing ourselves and was a great way to relieve some of the pressure that built up when living and working in small enclosed spaces, with the same people, for weeks on end. It should never be taken personally, but sometimes it could be a bit close to the bone.

Occasionally, they took turns to unstrap from the cockpit to walk down the back, stretch their legs and take a piss off the ramp. When we were flying, this was harder to do as piss tended to blow back up the ramp on to your kit, because of the wind and turbulence off the blades. On these occasions we brought out the 'piss tube', a small funnel attached to a tube that vented out of the side of the aircraft. Quite often the crewmen, when faced with a particularly annoying officer down the back who insisted on speaking to the pilots, would hand them the piss tube telling them: 'This is the standby intercom. Get your lips right into the microphone and shout loud. The pilots will hear you.' Hours of fun.

They sat turning and burning in the desert for four hours. A long, tedious, boring four hours sweating their nads off, wasting their lives, cutting their own personal holes in the ozone. It wasn't expected that it would take that long but the elders knew where all the minefields were so took the troops on a convoluted journey to make sure they weren't blown up.

The elders know more about what goes on in their area than us outsiders do. At the end of the day they live there. Most people know more about what is going on in their street than foreign outsiders do, which was essentially what we were. It was non-event day. If the troops had hit a landmine then they would have lost a truck full of troops and we would have been flying three aircraft into a shit storm. As it was they got the safe passage they were guaranteed. But I was done now. All I wanted was to get back home and off the Afghan beach.

Chapter 31

I had been forward at Bastion for the last ten days and it was dull as ditchwater. We had been called out on IRT shouts twice, early doors, but the last few days had been quieter than a quiet thing on a quiet day.

I had read all of my books and I had watched as many reruns of *Band of Brothers*, *Battlestar Galactica* and *Firefly* that I could stomach. We were short finals to getting home and that just made the seconds tick by even slower. We just sat around waiting to leave. I was done, I wanted to go home, to get back to my wife and kids, and have a few beers. There was nothing worse then being bored in theatre, other than being bored in theatre a week before you are due to leave.

I had been working out at the back of the IRT tent and then had dropped into the CP to see if there was a hint of any activity, but no, all quiet on the Western Front. There was nothing to do, but at the same time it was impossible to properly chill out because we knew that should the shit hit the fan, which it could quite easily do, then we would have to be on action stations, 0–60 in under a minute, so it made me a bit twitchy. I had lots of energy and nothing to channel it into.

When we had arrived Forward earlier in the week, one of the guys, who was notorious for being separated from his shit, had left his pistol in the IRT tent without realizing it and gone back to KAF. Having realized he had left his pistol he phoned Bastion to see if we had it. More fool him. We found the pistol lying on one of the cots. 'Ha ha!' I rubbed my hands in glee when I saw it lying there. 'We can have some fun with this.'

We carefully dismantled the pistol and dropped the separate parts into a plastic zip lock bag. We then filled a cool box with water, dropped the bag in, put it into the freezer and waited. Once it had fully frozen we popped out the block of ice like a giant ice cube. What we had created was a block of ice encasing numb nuts' pistol parts.

Twenty-four hours later a very flustered pistol-less chap had managed to jump on a cab to come Forward and collect his misplaced gun. He ran into the IRT tent. We were all hanging around trying to look casual.

'Have any of you lads seen my pistol?'

'Er, yeah, mate. Hang on a minute, I'll get it for you.'

'Eh?' He looked confused as I walked over to the fridge-freezer and opened the door.

'Here you go,' I said, pulled out the block and handed it to him.

Jonnie Porn started to snigger, which sent everyone in the tent off. We were all giggling like children as I handed him the massive pistol ice cube. Numb nuts did not see the funny side, and his face got redder and redder.

'For fuck's sake!' he said. He picked it up and smashed it hard on to the floor. The ice shattered and shards splintered everywhere. He grabbed the bag.

'You are fucking children, the lot of you! Wankers!' And clutching it he ran out of the tent to jump on his cab back to the Rear. We erupted into laughter. There were so many reasons why you needed to keep your shit in one pile in theatre, not least because aircrew have the mental age of eleven-year-olds! Freezing numb nuts' pistol had kept us entertained for about two minutes but once that was done and dusted the boredom set in again.

'I am so bored. Bored, bored, bored, bored, BORED!' I said to Evs, callsign Mafanwy, a tall gangly, young Welsh dude and a great mate, who was lying on his cot bed over the other side of the IRT tent.

'So am I! Why is it so fucking quiet?' said Evs

'Ramadan is coming up so the Teletubby activity has dropped off as they all go home for the holidays. Lucky bastards.'

'Shit, is it normally this quiet?'

'Normally they don't fight this hard for this long. It's been unusually nails this time round.'

Dan and Dave Hughes were playing PS2 in the chairs. Dave Hughes was a top bloke and sound operator. He was a young crewman, affectionately known as Diesel because he's got no spark.

'When are 27 coming up?' asked Dan.

'Tomorrow, mate, and they can't come soon enough.'

'Do you know who's coming up?'

'No, but I do know that most of 27 have seen all the HLS up here already as they were here six months ago, so the handover is going to be pretty short and sweet. I am going to show them the new stuff and then we'll get the fuck out of Dodge.'

'What's the plan to get back?' asked Dan.

'We are swapping out a cab so they'll come up. I'll show them the rota, and we'll do a quick handover then we'll jump into the cab they came up in and head back to KAF.'

'I am more than ready to get back now.'

'We all are.'

'In order to get us out of here ASAP, first thing tomorrow, Dan and Dave you make sure that all of the kit is taken to the cabs as soon as they get here. Evs, if you go with them then get the cab ready while I hand over. Does that sound like a plan?'

Everyone nodded in agreement. We didn't want to be hanging around unnecessarily.

The next morning 27 'C' Flight sent two crews Forward to

hand over the IRT and HRF. I handed over to 'Cuthie' Cuthbertson, a short northerner with black hair and a roundish face. He was a good lad and we had been on Prelim Ops together. 27 Squadron had done all of their training back at KAF and we were keen to get out of Bastion, so we kept the handover short and sweet. They had all been out before and it suited them. They knew what to do and they wanted to bed in as much as we wanted to check out. Once the handover was squared away one of the lads drove me over to the cab.

We fired up the engines and headed back to KAF. I kept myself in check – this was no time to let my eagerness to get home get in the way of my focus on the job. I wasn't going to let my guard down. It wasn't going to get any of us home if we flew into the ground because I wasn't focused on the flying.

Coming back to KAF is one time when you can't afford to be off the boil. By day and by night the airfield is a hive of air activity, with helicopters, fast jets and heavy transport aircraft all vying to get on the ground or away from the airfield as quickly as possible. KAF and its surrounding area was riddled with disgruntled locals and Taliban who would dearly have loved to shoot down an aircraft. Of all the missions we flew I always thought that the relatively simple job of landing at KAF was still one of the most dangerous aspects of flying operations in Afghanistan.

'Kandahar Tower, this is Footloose 24, field in sight, request join via Bravo sector at low level, for Foxtrot,' Evs transmitted on the radio.

'Footloose 24, Kandahar Tower, clear to join direct to Foxtrot.'

Nice. They were letting us straight in with no holding or rerouting. A sense of relief washed over us as we headed straight into the airfield.

'Kandahar Tower, Footloose 24, wilco and many thanks,' I added. It doesn't hurt to be polite to Air Traffic. I brought the cab in fast over the perimeter fence. We could see the faces of the guards in a watch tower as we flew over the top, dusting them down with our downwash. They grinned and waved. They kept us safe as we slept in our beds and we kept them safe when we were out flying, so the feelings were mutual.

I stood the cab on its tail to slow it down and then lowered the lever to drive the back wheels on to the ground.

'Down ten, eight, six, five, four, three, two . . . cock,' came the voice marshalling from the crewman in the front door as I held the back wheels off the ground for the last few feet, for a laugh to upset the crewman's rhythm. It was good sport. Finally we touched down and Evs taxied the cab back into our designated parking spot.

I wearily shut the cab down, pretty much by reflex and finally all was quiet. Shutting down the cab on our last flight on a det was a mixture of relief that so far you had 'got away with it' and wouldn't be flying on ops for a while, and sadness that this was the last time you would be strapping in to the beast and throwing yourself into the unknown, with the surges of adrenaline that accompany flying the Chinook in Afghanistan.

We drove back to the grots to find that 27 were well and truly in place. We had three days left in theatre, and so were put in these tiny transit grots. The room was the size of a postage stamp and furnished with four beds. I was sharing with Dan, Sam and another pilot called Redders. The room was piled high with us and our kit. It was cosy. We were living in each other's pockets, there was no flying and nothing to do apart from go to the gym, and we were just itching to get home. It became our mission to get back to Blighty ASAP. I was working hard on the movers to see how we could get out. I managed to snag some seats for Dan and me on a C17 flying out via Turkey. We boarded the aircraft and headed for home but we would be back soon enough.

Part 2

Chapter 32

30 June 2007

KAF was exactly as I'd left it. The same familiar mixture of part food, part military hardware. The six months had flashed by. In a heartbeat I was back in the dustbowl. For various reasons I had flown out alone. The old OC 'B' had moved on and I had a new one, Squadron Leader Nick Geary, managing my world. He was a strong leader, a great aviator and a real asset to the air force. It was a joy to work for him and we got on like a house on fire. He was my pocket battle buddy on account of his diminished stature, and my new det smoking compadre. His call sign was Bandit, a name loaded with danger and sex appeal. The reality, however, was that he was named after the dwarves in the film *Time Bandits*.

Nick picked me up from the terminal in the same old battered Land Rover that Digs had collected us in six months previously. I threw my stuff in the back and jumped in the front. I shook his hand.

'Hello, mate,' I said. 'I see that nothing has changed. Not even the transport!'

'Yes, indeed. Same shit, different day.'

'Talking of shit, I had forgotten how bad the smell of that shit pit is.' Fortunately, being back I was reminded in an instant.

'The good news is the new accommodation block is much further away from Poo Pond.'

'Oh yes, the Brit grots. What are they like?'

'We have only just moved in. Time will tell.'

As always I was looking forward to getting back into the thick of it again.

'So what's occurring? How was the handover?'

'Not so bad. The lads know what they are doing so the handover was relatively painless. I was told that we are to be involved in the biggest op that's ever been attempted to date, working with the Cloggies' – our affectionate name for the Dutch – 'US and Afghan National Army. They are calling it Operation Lastay Kulang,' he said (from the Pashtun word for 'pickaxe handle'). 'It will require air support for all three phases, the insert, the land part and the extraction. I have put Daffers as lead on the insert just to get the ball rolling and we are still working out who's going to go where and when. The scale is pretty epic. At the moment they are talking about fifteen or so aircraft.'

'I remember them mentioning it in the pre-deployment briefings but I didn't realize that it was going to be that big.'

As part of the work-up we had been given a heads-up to the ever-changing Intelligence picture in theatre. These briefs were essential for getting us ready, even though I was coming back for the third time. Each deployment was different. We had learnt not to assume that it was going to be the same as the last one.

Lastay was going to make the Sangin RIP seem like a walk in the park in terms of organization and logistics. I could feel another QHTI lob-on growing in my pants.

'When does it go?'

'Next week.'

'Really? Shit.'

'Daffers is a good lad. He'll get a grip on it,' he said.

'Am I playing then?'

'No, mate, you are going Forward the day after tomorrow, with Steve Davies.'

My QHTI lob-on drooped, knowing that I wasn't going to be involved in the planning for the insert.

As we drove to the grots at the main base, it seemed busier, particularly as it was late evening.

'Is it me or are there more people here?'

'Lorimer has increased the numbers to around 6000.'

'Shit, it's doubled in size!'

The Army's 12 Mechanized Brigade, which included the Royal Anglians and the Grenadier Guards, had taken control of the British side of military operations in Helmand under the leadership of Brigadier John Lorimer. They had relieved the Royal Marines from 3 Commando Brigade.

The arsenal had increased as well. There was now more heavy equipment, namely Warrior infantry fighting vehicles (IFVs), Mastiff protected vehicles and Guided Multiple Launch Rocket System (GMLRS) launchers. The war machine was growing and turning over at a rapid pace.

'As you know, mate, 12 Mech have taken over and we're at the beginning of a new phase of multinational operations that are in the early stages of implementation. When you were here before we worked more independently as a country, but now it seems there is more effort to integrate the ISAF nations and for us to work cohesively.'

'Should be interesting. What's been done so far?'

'At the moment we are in the middle of Op Achilles. It's a pretty huge op with little sub-ops going on within it. There are five and a half thousand troops, ISAF and Afghans, occupying all the strategic points of Northern Helmand, all the DCs, and the Kajaki Dam.'

'What's happened on the satellite ops then?'

'Within Achilles there is Op Kryptonite which happened

last Feb. The Cloggies, the Brits and the ANA cleared an area from which the Taliban had been mortaring. One of the heads was taken out in an airstrike, a chap called Mullah Manan. He was a nasty bastard and up there in the top tiers of the Taliban hierarchy.'

I listened as Nick updated me on the ops SITREP. Operation Silver had happened in early April. A 1000-strong group of US Paras from the 82 Airborne Division, along with a group of 250 Royal Marines, had completely cleared Sangin.

Later that month there had been Operation Silicon, which had been conducted mainly by ANA troops, supported by British units. They had totally cleared the Taliban out of Gereshk and the lower Sangin Valley. It was within this op, in mid-May, that the Taliban head of sheds, Mullah Dadullah Lang, was killed in Gereshk district, by a UK security forces and ANA collaborative operation. This particular bad boy had been the leader of the Taliban military in Helmand, so his loss would be keenly felt and retaliation was expected.

Op Silicon was still raging on and there was fierce fighting with the Taliban at Sangin, as the ANA and Brits tried to establish a more permanent presence in the area. Achilles was scheduled to finish in a few days' time and then Operation Lastay Kulang would kick in, the very same morning. Daffers had six days left to get his shit in one pile.

We pulled up outside the new grots. They were bespoke fifteen-room, air-conditioned, massive mobile homes. There was a shower block and the rooms were laid out in a regimented military style. I grabbed my kit and dumped it in the corridor.

'Where am I kipping?'

'Sorry, mate, as you turned up late you are in the TV room.'

We had set up one of the rooms as a communal area. I was to sleep in here. It had its advantages, because I was in there on my own, but I didn't get much peace and quiet as everybody would use it to watch TV. It didn't matter because I was going Forward anyway.

After I had dumped my kit by my bunk I was stood outside having a fag in rig when an Army Major I didn't recognize walked past me and entered our grots. The grots was our home for the next eight weeks and although we were all living together in quite close quarters and didn't get much privacy among our fellow flight members, we didn't take kindly to strangers walking in and around our private space.

A few minutes later the Army Major on the way out walks past me again.

'Excuse me, mate, what do you think you are doing?'

He stopped and looked at me as if to say: '*What the fuck has it got to do with you?*'

'Have you just taken a dump in our heads?'

'Yeah!'

'Er, where do you work?'

'In the NSE.'

This is the National Support Element, an administrative outfit that supports the Brit ops in KAF, aka a K-Span full of blunties (admin staff)!

'Well, why didn't you take a dump over there?'

'Because they are porta-potties.'

'Ok, why don't you walk back to your block?'

'Because they are 150 yards over that way and I was taken short.'

'Ok, well that block next to us is closer than our block so why didn't you poo in there?'

'Well, you know why, because that is for VIPs.'

'So you won't poo in the VIP block because you'll get caught but you are quite happy to take a dump in our block?'

'Well, I don't really see your problem.'

'Well I do, mate, these grots are our home and the heads are at the end of the corridor. How would you like it if some univited random bloke walked through your house and took a dump in your heads? In future, go and ablute elsewhere!'

He looked at me as if to say: '*Yeah, whatever, prick*' and walked off.

Daffers walked up to the grots while I was still seething over another fag.

'Hello, mate.' I reached out and shook his hand. 'Did you see that prick that's walking away? He's a blunty that works in the NSE. It looks like they are using our heads to take a crap in because they have porta-potties in their K-Span.'

'Cheeky bastard!'

'Too fucking right! Do you want to grab a coffee?'

'Cracking.'

Daffers and I walked through the grots and out into the area known as Cambridge Lines, the Brit accommodation area, stretched over two acres of prime Kandahar real estate.

'Mate, we have our own coffee shop now?' It was called Pinchettos and was run by the NAAFI.

'Yeah and we have a Timmy Hortons, up on the board-walk.'

'What's Timmy Hortons?'

'It's a Canadian chain, specializing in doughnuts.'

'Do we still have the Green Bean?'

'It's still there but not as busy as it used to be considering the new competition.'

Pinchettos looked just like a high street, glass-fronted shop. Inside there were tables, stools and couches at the back. Widescreen TV adorned the walls. We walked up to the counter and were greeted by a female British NAAFI employee in her mid-thirties, one of the many civilian contractors now working at Kandahar Airbase. The menu included the full range of poncy coffee and cake combinations. It was surreal.

'Can I help you?'

'What do you want, mate?'

'I'll take a hot chocolate, otherwise I won't be able to get my head down.'

Daffers ordered and carried our drinks over to the comfy couches to catch up.

'So you've got Lastay. It looks pretty involved.'

Daffers explained to me that the US were leading the op but the aim was to build on the achievements of previous operations in the Sangin Valley, mainly Achilles, which was now drawing to a close. The whole operation had been at the request of the Afghan government and we were working in collaboration with the Afghan National Security Forces (ANSF).

Strategically, Lastay Kulang was extremely important. It was focusing on the area between Sangin and Kajaki, known as the Upper Sangin Valley. Geographically, the Sangin Valley from Kajaki to the south is one of the major trade routes for that part of Afghanistan. It travels down towards Pakistan and straight through the Chagai Hills on the border, some 300 miles away in the south – a notorious drug-manufacturing region. It follows the Helmand River for most of its course before turning into relatively inhospitable desert for the last hundred miles or so, then crosses the generally unguarded border into Pakistan by the frontier town of Bahramchah, host to regular drugs and arms bazaars.

The Taliban's main focus was to maintain the supply routes for drugs going back and forth to Pakistan, the various heroin-processing factories and the supply of people and arms coming to and from Pakistan. The Taliban are hugely funded by the production of heroin as well as other elements. They also use this route to deploy fresh blood and rested fighting forces from the south, bringing them up through Garmisir, Lashkar Gah and into Sangin and Kajaki.

Lastay Kulang was to target the key Taliban stronghold area south of Kajaki and north of Sangin, with the sole aim of kicking them out. Altogether, 2000 ISAF and ANSF personnel were to take part in the operation, 1000 of them British forces, from the Royal Anglians and some from the the Grenadier Guards, who were working very closely with ANSF. The Helmand Task Force also included Danish and Estonian troops. The main aviation assault comprised Taskforce Fury, the American 82nd Airborne Division, or 'All American Division'.

The plan was to insert the All American Division and they would push the Taliban back out into other less strategically key areas. Then they would work quickly and sympathetically with the local population to make sure the Taliban didn't immediately infiltrate back in.

All of the operations that were being conducted had to be in harmony with the aims and goals of Karzai's government. The governor of Helmand had his own missions and goals in the area in support of his people. Having to deal with all the inter-tribal rivalries, as well as the Taliban infiltration, made it a complex process, but ultimately it was our job to support the government of Afghanistan, via the governor of Helmand.

Most of the people of Helmand live along the river leading up to the Kajaki hydroelectric dam, which could provide power to Helmand Province if it worked. The long game was to fix the Kajaki Dam, but this wasn't possible until conditions had been stabilized along the supply route.

This could only be achieved in stages owing to the asymmetric nature of the conflict. The war was asymmetric because the methods and tactics deployed by the Taliban did not match ours. They were unbalanced and unconventional and did not confront us on equal terms or with 'symmetry'.

The Taliban had infiltrated many if not most of the villages in the Upper Sangin Valley. They would commandeer a civilian's house and say: 'I am going to live in your house for a week.' The locals then had no choice but to allow them to live among them. They said: 'I will offer you protection in return for your looking after me,' but in reality the Taliban were only out for their own interests and not those of the local population.

But for us, strategically it wasn't a case of starting south and marching north until the area was cleared. This was not practical and there were not the assets or resources to enable the allied forces to undertake such a tactic on the scale required. There was also no way that anyone would have accepted the cost to civilian life or damage to the infrastructure that would have been incurred in order to achieve our mission in such a manner. We couldn't simply carpet-bomb the place because the Taliban were living in among the civilian population.

The Sangin Valley was littered with compounds, small farms, houses and storage areas. The cultivated areas either side of the river were known as the Green Zone. From the air the compounds could be seen very clearly, as could any activity around them, but the troops on the ground often couldn't see what was happening in a compound as they approached it. The Taliban could be almost anywhere and the only way their locations could be established was by surveillance over a long period of time, examining their pattern of life and establishing where they moved and rested.

Every mission had to be carefully considered, planned and arranged to be targeted specifically. It was a surgical operation that was executed in as precise a fashion as it could be. Each operation had a time limit, so the idea was to go in and achieve as much as possible in as short a time as possible while setting the conditions to stop the immediate return of the Taliban.

This was achieved by the UK and allied insertions going in and basically kicking the Taliban out. The Taliban would only fight for so long and then they tended to retreat to a safe haven, melting away into the population, often leaving weapons behind for fear of being stopped and searched, although every Afghan male above the age of ten seems to own an AK-47.

We called it an 'Afghan low-flying noise complaint' when they took a pot shot at our helicopters, like tin cans on a fence, on the way past their farms and fields. Luckily they are usually pretty poor shots.

The allies would then follow up this initial offensive phase with quick-impact projects such as helping get the school reopened or fixing the road. They would often eat with the local elders in a Shura (council) to find out what their needs were, before making that quick impact so that the villagers would think: *This is great, we don't want the Taliban back, this is what we want – peace, security and stability, ultimately governed by the Afghan National Security Forces.*

This then sets the conditions for the locals not to want the enemy back in, as the Taliban are seemingly offering them less than the allied forces. We were then pushing the Taliban around the area, up, down, back and forth, shepherding them into areas that suited us rather than allowing them to be in areas that didn't.

'How's the planning going?' I asked Daffers.

'The difficulty is, mate, there are so many troops to insert that it's going to be impossible to achieve it in one wave. It's looking like it could be two, even three waves.'

All of the waves had to be inserted into a tightly confined

area, and it was crucial to the operation that all troops on the ground were put in as quickly and as closely as possible. The only way that this could be achieved safely was to have three separate routings in and out on every occasion, flying over separate pieces of terrain except for the final stages, when we would be running into a landing site.

'Who are you working with?'

'The American aviation outfit are running it, Taskforce Corsair, and we have also got the Cloggies with us.'

The waves would consist of combinations of formations of two British, two Dutch and three American Chinooks, plus six Black Hawks and six Apaches for top cover.

'We are restricted in our ways in and out. The LS is a large lozenge-shaped bit of desert in the Upper Sangin Valley. We have got to put troops in at the river and drop their artillery on the other side.'

'What are the Anglians doing?'

'They are moving out of Bastion, on the ground, in order to put a blocking position to the south, north of Sangin.'

'Who is doing it with you?'

'I have Tweedle Dee and the other cab is Zippy and Simmo.'

Tweedle Dee was a top bloke but so named because of his rotund appearance. Zippy talked a lot, and Simmo's real name was Rich Simpson, so his nickname was in line with the military's classically imaginative surname abbreviation.

'How are you getting on with our American friends?'

I had an understanding of the challenges Daffers faced with the Americans from my time with the US Marine Corps. They have a noticeably different take on how to do business.

'The Americans have got facilities and stuff that we can only dream about. Kit that enables them to come up with this amazing plan to co-ordinate all the aircraft in. If the

Brits had been planning it we would have done it on the back of a fag packet.'

'Right then, that's that, follow me, don't crash!'

I laughed, but Daffers' humour rang true. We Brits seem cavalier but we are not. We are consummate professionals. We have to make do and although it's not 'bodge it and scarper' we have to work with the kit we have been given. And ultimately, it might be said, we are better aviators for it, partly because of how well we handle what we do with what we have got. It certainly gives us a fresh perspective on life and maybe we just don't take ourselves quite so seriously as our American friends because of it.

We finished our drinks, headed back to the grots and continued our conversation in my new bedroom, aka the TV room.

The planning process not only involved the aircrews but also the ground troop commanders. It was all taking place in the briefing room of Taskforce Corsair.

Daffers gushed: 'They have this wonderful computer, I can't remember its name.'

'Falcon View?'

'Yeah, that's it. The American briefing officer said to me: "Chinook one from the Brits, this is your position as you fly in and this is what you will see as you fly in." I looked at the computer monitor and there was a 3D graphical simulation of our route into the Upper Sangin Valley. It showed me a 3D graphic of the aircraft and what the terrain would look like as we flew up the valley to the landing site. We were gobsmacked. "Wow, that's incredible," I said. I was really impressed. And then he said: "And for the ramp crewman, this is your view." He moved the mouse and then the crewman's view appeared. It was fucking brilliant. Then this American pilot walked past me with his top bollocks electronic kneeboard. I turned to JB, and said: "Shit! Did you see that electronic kneeboard thingy. Why can't we have one of them. I want what he's got.'"

Daffers talked enthusiastically about the planning process. We were like kids in a candy shop. The Americans had all the gear and we were the poor relations that had been invited over for tea and a bit of a play with the Gucci toys.

From what Daffers was saying the amount of effort that the Americans were putting into the planning was phenomenal but came as no surprise. Every time I had worked with the Americans they went into this level of detail.

I was knackered and needed to get my head down. I had a day of admin ahead of me before going Forward to Bastion. I told Daffers we would catch up the next day. 'Before you go, mate, what's the score with scran?'

'We now have a Brit DFac over the road. It's distinctly average.'

'What's the latest time to grab breakfast?'

'Around 8.30 is pushing it.'

'Have we still got the PoshFac.'

'Yeah, but we are not meant to use it. All the blunties from JHF(A) seem to be using it though.'

'Typical!' Some things never change.

Chapter 33

The next evening I was sat in the TV room and Daffers walked in. I was not in the greatest of moods. The inevitable back in theatre admin had got me down.

'Good first day at school? Did you play nicely with the other children?'

'Don't! Niff naf and trivia does my head in. Today, mate, I had a shit breakfast at the new DFac. Thanks for the heads-up – yes, was very average. I have zeroed my weapon, received my Int brief, had the update on the HLS, picked up my morphine and squared flying kit. How about you?'

'I have just finished the Lastay AMB' – the Air Mission Brief. 'I never cease to be impressed by our American friends. As we walked through the door we were given briefing sheets with an outline of what we were going to do with this Gucci pack of all the planning that we needed for the op. It had all the frequencies, call signs, routes, waypoints, code words, the whole kaboodle.'

'Yeah, it's called a Smart Pack, they used them all the time on Cobras.'

He explained that in total there were about 250 waypoints and, of course, the Americans had a little machine that squirted all the waypoints into their computers, but not us Brits. Daffers was laughing because Tweedle Dee was going to have to sit in the aircraft for three hours manually inputting over 250 waypoints into it. It was going to be really tedious.

Tweedle Dee had pointed out to him: 'If the aircraft goes U/S and we have to get a spare, we are fucked and we won't know where the fuck we are going.' There would be no time to manually reinput 250 waypoints.

'We were sat listening to the brief and there was one point when JB and I sat there shaking our heads fiercely.'

JB was flying the second Chinook. He like Daffers was a seasoned pilot.

'"Err, is there a problem?" said the American briefing officer.

'"Yes, yes, there is. You have three waves of aircraft going into a very tight space and you have all three waves converging at this point. If one aircraft overshoots we are going to have the biggest mid-air collision in the history of aviation. We really don't like the fact that we are all flying towards each other,"' Daffers explained. '"What we would really like to do just before we get there is actually all be on a similar track just in case one aircraft can't get into the landing site and has to go around. Then that aircraft is not going to collide with some-body trying to lift out of the LS. It's dangerous."

'"Yes, right. We see where you are coming from," said the American briefing officer.

'The LS was only 800 metres from top to bottom and was expected to be able to accommodate three Chinooks and two Black Hawks in the American formation, two Chinooks and two Black Hawks in the British formation, and two Chinooks and two Black Hawks in the Dutch formation. A total of thirteen aircraft were coming into the LS at once. The Ameri-cans had given us a two-minute split between packets arriving and departing. The aim was that we would get in, drop the troops off and then get out again before the next formation landed. But on the egress they had all us diverging. JB and I both looked at each other and raised our eyebrows. We had studied in intri-cate detail all the photographs of the LS. Those Americans have got the latest satellite imagery. It makes a mockery of our 1993

Polaroids left over from the mujahideen and the best that Google Earth could offer,' laughed Daffers.

'When I looked at it, it was pretty clear that it was going to be tight to get us all in. We have two Black Hawks with us as well as another Chinook. I'd said to the US Warrant Officer that was flying the Black Hawk in my formation: "You need to make sure that you push right into the landing site to give us room."

'"Yeah, no problem, sir, I can do that for you."

'"When we come out of that site we don't want to be hanging around because of the threat of the .50," I'd said.

'"I hear you loud and clear, sir!" he said.

'I just hope that he did because we are going to be in a whole world of trouble if he can't get his aircraft in exactly where it needs to be. One of the reasons they wanted us to go up there was because there had been some heavy-machine-gun fire on the recce troops in the mountains which they hadn't been able to deal with.

'There was a lone Taliban shooter with a .50 weapon up there that was killing these guys so they wanted to flush him out. He had been pinning the troops down right on the track line on the egress. But when I studied the planning we were the cab which was flying right over this bloke's position!

'I was like, "Err, excuse me. Why are we having to fly over that heavy machine gun. That is not a good plan!" The reality was we had drawn the short straw and that was that, so we're flying over the shooter then – brilliant!'

'When are you tying it all up?'

'There is an ROC drill the day after tomorrow.'

The Rehearsal of Concept brief was the final walk-through, where everyone stood as a crew in formation and walked through the entire insert.

'I am going Forward tomorrow to pick up IRT with Steve Davies, Redders and Tommo.'

'Mate, now I need to go and blow off some steam. I am off down the arm farm for a full body buff.'

'Freak!'

With that I grabbed my gym bag and headed to the gym for a dose of anger management. After the gym I went back to the grots to take a shower. In the bathroom of the grots there was a line of showers at the entrance, alongside a line of traps and then opposite the traps were the sinks. Redders and Tony were at the sinks brushing their teeth. I was wrapped in my towel, having just got out of the shower, brushing my teeth next to them, when in walked a bespectacled, spindly Army Major. He walked over to one of the stalls.

'Ahem, what are you doing?'

'I am going to the toilet.'

'Er, not in here you're not. Do you work at NSE?'

'Yeah.'

'Well, you are not taking a dump here. These heads are only scaled for the number of people in this block. This isn't the annexe to the NSE toilet so get out.'

Redders and Tony are stood smirking at the sinks passing each other looks with raised eyebrows.

'Who are you?'

Right, you bastard, I am going to have some fun with this, I thought.

'I am Corporal Hammond.'

'And I want to know who your line manager is and I want to talk to him. You will not speak to me like this as I am a Major.'

I walked towards him, still wrapped in a towel and said: 'Nah, I am only joking, my name is Major Mark Hammond, I am in the Royal Marines and I am my own line manager. But you are not taking a dump in our heads so get out.'

With that the bespectacled Army Major turned on his heel

and just as he was leaving said: 'I am not happy about this. You'll be hearing from me.'

I didn't think any more about it. I never discovered where he dropped his log but he went from our grots to his CO, a naval Captain, and complained that he had been bullied by a Royal Marine officer in the toilets and that he would like to file a formal complaint because he felt threatened.

Chapter 34

The IRT grots had improved in the last six months. Where there had been cots, there were metal-framed beds with sprung mattresses, duvets and pillows. The collapsible camping chairs had been replaced with Ikea armchairs, provided by the welfare fund. If it was indicative of how long they were planning on keeping British forces in Afghanistan this was to be no short holiday. We were in for the long haul.

We had been in situ less than twenty-four hours when we got our first IRT shout. Redders, Steve Davies, Tommo and I were watching TV. Tommo had replaced Mr Mack as the Master Aircrewman. I had gone through the OCF with him, an ex-para and air engineer on Hercs though that would be difficult to tell from his rotund body. He was a great operator and a benefit to any crew. Surfer Dude Redders, originally from Cornwall, sported a non-regulation haircut. It drove me insane. I kept threatening to shave it off in his sleep. He was my new gym buddy. Steve was with us in 06. He was originally on an Army Air Corps exchange but transferred to the RAF permanently. They were good lads and I knew we had a swept-up crew.

We had already had an interesting meeting with the new AH guys as we sussed out the lie of the land.

'Things have changed here. We don't seem to have that close working relationship with the battle group any more.'

'We don't go into the battle group CP for IRT shouts now.'

'OC Forward seems like an interesting character. I sense there will be some conflict there.'

Inevitably, the AH squadron had changed and so had the dynamics. It felt like it was them and us, rather than the integrated team we had in 06. As usual, OC Forward was the AH squadron boss, but I didn't know him of old and his personality was diametrically opposed to that of McGinty. He seemed to me to be everything I most disliked in an Army Air Corps officer, an ambitious promotion thruster who just happened to also be a pilot. I could see that we would be butting heads.

The Royal Anglians were the incumbent Bastion battle group but the relationship was far more divorced then it had been with 3 Para and 42 Commando.

Redders said to me: 'Mate, did you know that we had Ross Kemp down the back on the way over.'

'You are shitting me.'

'He was sat down the back, with a cameraman. I didn't recognize him at first. He's a lot shorter than you think.'

'What's he doing here?'

'I have no idea. Probably a programme on the gangs of Kajaki.'

There was field telephone in the corner of the IRT tent. The radio comms of 06 were long gone. The phone rang twice.

I picked up the phone. 'IRT.'

'Sir, we have a nine-liner.' This meant there was a casualty. A nine-liner was the format made up of nine lines indicating the position, number and category of casualties.

I put the phone down.

'Gents, we have got a shout.'

The well-oiled IRT machine kicked in. Steve, Tommo and Redders legged it to the cab to flash it up. I ran over to the JHF(A) CP to get the detail and the Green Brain.

'What have we got?' I asked the Chief of Staff, a short, squat Major who had been commissioned from the ranks and was a good bloke.

'One of the Vikings has hit a mine.'

The Royal Anglians were conducting the prelim moves prior to the air assault for Lastay. The guys had left Bastion in a convoy comprising trucks, armoured vehicles and the Vikings, tracked vehicles bought from the Royal Marines and replacing the ageing BV206. The Vikings were designed for Norway but had proved their mettle in the desert when 3 Commando were in theatre, and the decision was taken to keep them out in support of 12 Mec Brigade. The Anglians had been moving up the eastern desert to establish their blocking positions.

'There are two injuries and one of them is pretty bad.'

'Ok, what's the grid?'

This was passed to me and I checked it out on a map. I briefed with the AH escort. It was out in the desert so safer than going into one of the DCs. We still needed to keep our wits about us. We didn't know if it was a minefield or an isolated mine as we had to land on to pick up the casualties.

'Have they cleared an area?'

'They are doing that now.'

We needed to make sure that the area was clear so that we could safely land on. We couldn't escape that fact that there were known mines in the area.

Tommo came back to pick me up and we arrived at the already turning and burning IRT cab. Everyone was on the back waiting. I strapped in and within seconds we were ready to lift. I put the grid reference of the pick-up point into the CINS, our onboard navigation system, and we routed at height to the area, accompanied by our AH escort.

The AH cleared us into the HLS.

'Visual the smoke.'

As we dropped to low level, the Anglians on the ground had released a smoke grenade denoting the cleared area for us to land on.

'Roger, visual.'

We came into the gate. Redders talked us down to wheels on. As the dust cleared I could see in my one o'clock the shattered remnants of the exploded vehicle. The Viking is made up of two parts: the armoured tractor at the front tows an armoured box, within which troops sit at the rear. The whole aft of this box had been blown to smithereens. It didn't look pretty.

'Casualties coming onboard.'

The casualties were loaded.

'Ramp up, clear above and behind,' said Tommo.

We were engulfed in dust as we lifted out and routed back to Bastion.

'How's he doing?'

'He looks in a pretty bad way.'

'We are only ten minutes out.'

'Crowbar, Morphine 25, routing Bastion, request picture.'

'Morphine 25, Crowbar, picture clean.'

Crowbar was an air space management service, controlled from Bastion. It had an enormous radar which kept tabs on all helicopters and jets in the Helmand area. It provided traffic information and updates on any active Restricted Operating Zones (ROZs).

'We will need Plan Two on return.'

This was asking for the large radar to be turned off so that we could land at the hospital HLS without the aircraft systems being affected by the output from the radar.

I switched to Bastion Tower. 'Bastion Tower, this is Morphine 25, inbound for Nightingale. Request Plan Two.'

'Tower, Morphine 25 cleared land Nightingale, you have Plan Two.'

We landed at Nightingale.

'Ramp down,' said Tommo. 'Ambulance approaching.'

Now that theatre had scaled up to double the troops on the ground, the old days of the RSM and CO waiting at the HLS were a dim and distant memory.

The casualties were offloaded and bluelighted to the battlefield hospital. We repositioned to the spots and shut down.

I headed over to the CP to drop back the Green Brain. In there was Captain Nick English. He was an AH pilot who had been out with us in 06. A tall, gangly officer with a sharp wit, he gave as good as he got.

'Hello, mate, I didn't realize you were back out.'

'Yeah, they transferred me from 9 Regiment to 3 Regiment to help out.'

'They seem like a different bunch.'

'Yeah, they are a little wet behind the ears.'

'How come I haven't seen you until now?'

'I have been back at main doing the ROC drill for Lastay.'

'That must have been fun for you. Was that your first one?'

'It was little tortuous.'

'Hoofing, aren't they?'

'Where the fuck did that hangar come from? It was huge. I didn't even know it existed. And the level of detail they put into the model was something else.'

The ROC drill had taken place in a huge hangar at KAF. There was a physical representation of the route in, the objective area and the route out laid out on the ground, the size of approximately two tennis courts and filling two thirds of the hangar. All the routes that they were going to fly were laid out with coloured tape. The detail included the boundaries and where to make radio calls.

'Are you still on for tomorrow night?'

'Yes, mate, we are launching out of here and I have come back to relay the plan to everyone back here.'

'Good luck. I hope it goes well. We are on IRT for the insert.'

Chapter 35

A runner from the CP stuck his head into the IRT tent.

'Sir, we have a Chinook down.'

'Fucking hell.'

Redders, Steve, Tommo and I froze for a second in disbelief.

'Was it one of ours?'

'I don't know any more, sir.'

'Right, I am coming across to the CP.'

I looked at the guys. 'We can't all go over because it'll be busy in there. I'll give you a bell when I have got more detail. Steve, are you coming?'

Steve and I headed straight to the CP. It was surprisingly quiet. I went straight up to the Chief of Staff.

'Do we know what's going on?'

'It's a bit confused but we don't think it's one of ours.'

I looked at Steve.

'I am going to call Nick. He's in JHF(A) main ops room. He'll know the crack.'

I picked up the phone and dialled KAF on a secure line. One of the signallers picked up the phone.

'It's Major Hammond, can I speak to Squadron Leader Geary?'

'Wait one, sir.'

'Squadron Leader Geary.'

'Nick, it's Mark, is it one of ours?' I asked him.

'No, mate, it's not one of ours. It's an American cab.'

'Shit! Have we got any details?'

'It's all a bit sketchy at the moment. Looks like it might have a taken a hit on the way out.'

'Are there any survivors?'

'Doesn't look like it. But it was after they had dropped off the troops. There would have only been four or five onboard, not that it makes it any easier.'

'Mate, based on experience from the Nimrod going down last time we were here you'd do well to get hold of somebody back at Odiham sharpish and tell them now that it's not one of ours before the press speculation goes mental,' I said.

'Yeah, I'll do that as soon as I can. We have to crack on here.'

'What? The lads are going back in?'

'We have to because we have already committed the first wave.'

'Fucking hell, mate!'

'Mark, I have to go.' With that Nick hung up the phone.

I walked back over to the IRT tent. The guys were waiting expectantly,

'It's not one of ours,' I told them.

'It looks like it was an American cab. To be honest it's a little confusing. We don't know how many casualties or what has happened.'

'How many are dead?'

'I don't know.'

'How did it happen?'

'I don't know.'

'Was it enemy fire?'

'I don't know.'

They soon realized that I didn't have the answers. The room fell quiet.

'What's happening now? Have they scrubbed it?'

'No, they are cracking on.'

'They have just lost a cab and they are sending them back in?'

'They have to have boots on the ground, they need to get the rest of the lads in.'

'Fuck, the lads must be shitting themselves.'

We were stunned. We were all questioning our own mortality. We conveniently dismiss the fact that we do a difficult job in dangerous circumstances. Our lives hang daily by a thread which we thought was as thick as a three-metre strop but in reality was as fragile as a cobweb. It reminded me of the IRA statement to Maggie Thatcher after the Brighton bombings: 'You have to be lucky every day, we just have to be lucky once.'

We needed to be in crew rest to cover the IRT. I had to get my head down. I lay on my bed trying to sleep, thinking about the lads flying back into an area where a cab had just been shot down. Was I going to lose some more of my friends tonight?

Chapter 36

Forward at KAF, the plan for the Lastay Kulang insert was set but had changed at the eleventh hour and they had decided to have only one Brit Chinook on the first insert, which was Daffers with Tweedle Dee in his left-hand seat. Down the back were two great crewmen, Andy Bennett on the number one and Craig Fairbrother on the number two.

The HALS at Kandahar were mad with activity as the troops mustered, and the ground crews and engineers did the last-minute checks. It was impressive to see seventeen aircraft, US, Brit and Dutch, Chinook and spare Apaches and Black Hawks, resting on the pan, their rotors bowed, waiting to be fired into life.

Daffers and his crew boarded their cab and began their pre-start checks. There was a frisson of tension in the air, especially as they knew they were going to be flying over the .50 calibre heavy machine gun which had been pinning the Americans down over the previous five days. They hadn't been able to destroy the gunner so he was still active and dangerous. He was a real threat. It was a big op and 18 Squadron hadn't been in theatre long. This was to be their big hello to the det.

Daffers put his helmet on and lowered his gogs over his eyes.

He turned to Tweedle Dee and said: 'It's going to be a long night with these fuckers on.'

They were flying under US SOPs, which meant they were going to be flying for six hours solidly at mid-level. They were apprehensive about it. To spend such a long time at mid-level was, in their opinion, unnecessary and presented risks they didn't even want to think about. But that was the plan. They'd live with it.

They ignited the engines and the aircraft burst into life. Once the troops were loaded, they taxied over and lifted into the Afghan darkness. Dotted ahead of them and enhanced by the green hue of their NVGs they could see the burning heat of the engines of the Chinooks in formation. The Black Hawks were absorbed into nothing-ness by the pitch black of the night sky.

As they were coming into the gate approaching the LS, Daffers said to Tweedle Dee over the intercom: 'Let's see if that Black Hawk guy is as good as his word. He needs to push in really tight.'

'There he goes,' said Tweedle Dee. 'C'mon, mate, in a bit, in a bit more.'

'No way!' screamed Daffers. 'He's fucking gone short! Cheers, mate.'

Daffers was thinking: *If I overshoot I am going to land on top of those Black Hawks and I have got nowhere to go.*

The LS was very tight for the Black Hawk to land on. Daffers had known that the pilot had to push right in to make enough room. He didn't. He landed so short of it that they only had one place to go and when he put the aircraft down his rotors were just 7ft above a compound wall. It was a good job the other cab wasn't with him because they definitely would have had nowhere to go.

Daffers brought the aircraft to the ground and was engulfed by the dust cloud. As the cloud cleared he could see the American Chinooks to the left and was impressed by the brilliance of the Yanks' drills. They were ramp down and all of them were off in twenty seconds. Slick and streamlined and it meant that Daffers was ready to go but he had to sit there and wait.

He said to Tweedle Dee: 'Can you see that we have just got rid of twenty blokes in a heartbeat. What the fuck are those monkeys playing at?'

The Yanks had called TOWRICO ('take off when ready in chalk order') so they were sat on the LS watching six blokes fannying about getting off the Black Hawk.

'C'mon, c'mon, c'mon – we don't want to be sat here, we want to get going!' It was beginning to piss him off.

He looked at Tweedle Dee. 'When we leave here, your job is to make sure that I don't fly into the ground in front of us. We will be no higher than a hundred feet and we will be at max chat, to warp speed to get the fuck out of here.'

He called Craig on the number two: 'If that .50 gets one round off at us, you make sure, mate, that it is the last thing he ever fucking does.'

They were all geared up for the exit. When the Black Hawks eventually lifted they went off too quickly and went much higher than Daffers had expected. He thought: *Brilliant, they are smaller than us and now we are the biggest target. We are going to stay really low and really fast.*

As he lifted out he saw that an American Chinook was still on the landing site and thought: *Shit, they should have gone ages ago.*

They were having difficulties getting a quad and trailer off. Daffers cleared the river and the two Black Hawk pilots in front of them, unaccustomed to flying at night in tight formation, slowed down.

'No, don't slow down, don't slow down. Oh fucking hell!' he screamed over the intercom. He'd had better days.

They made first turning point on the route back to KAF, and Andy, the crewman on the number one on the ramp, called over the intercom: 'Huge explosion, half a mile, eight o'clock.'

'What does it look like?' asked Daffers, trying to picture the view and make sense of it.

'I think they have just JDAM-ed something cos it came in at an angle.'

The angry fireball lit up the night sky. They didn't know it at the time but Andy had just witnessed Flipper 75 going up in flames. A Chinook down.

They were halfway back to KAF just past the next turning point, going back to reload more troops, when Andy piped up. 'We have got another aircraft coming up behind us.'

'What, there shouldn't be anyone behind us.'

Daffers was confused. 'The Americans are going back in a different direction and so are the Dutch. So there definitely *shouldn't* be anyone behind us.'

The aircraft following them was the other American Chinook. The rest of the package had gone. He had been waiting for Flipper 75 to go but 75 had been shot down. Now the remaining American Chinook didn't know where to go. He had picked out the glow of the exhausts from Daffers' cab on the goggles and had decided to follow them.

'What the fuck is he doing?' said Daffers.

The two Chinooks, Daffers, and the following American continued the transit back to KAF and refuelled. But while they repositioned on the dispersal to reload they were swamped by engineers. 'What are they doing?' Daffers asked over the intercom. 'Why are they here?'

The message had got back to the ground crew that a Chinook was down and they were keen to make sure that it wasn't a Brit. At that stage nobody knew whether it was Dutch, American or Brit. And Daffers and his crew were still in the dark. Strangely, though, the engineers didn't tell them why they had come to check the aircraft. They just left Daffers and his crew to puzzle over why they were all hanging around.

They waited for the next lot of troops to be loaded. The landing sites were located at different places at KAF, so each set of troops knew where to go to board each aircraft. The American cabs were further down the dispersal, and Daffers was in the middle, with the Cloggies to the right. Watching the troops load seemed surreal.

The next lot of the 82nd Airborne marched over to the cab, loaded with Bergens. It was the military version of holiday reps shepherding holidaymakers to different coaches at the destination airport, but more organized and better executed.

They were ready to launch on the second. Zippy and his crew had joined them for this part. Daffers was handed a map and on it was a marker that said 'downed aircraft'.

'Look at this.' He pointed to the map.

'That cab must have gone unserviceable. He must have had a technical problem. That explains why the other Chinook picked up our tail.'

They were still none the wiser. Daffers thought the aircraft had gone U/S at the LS because nothing had come over the radios and nothing seemed different apart from the random Chinook following them. Daffers and his crew were then joined by another Black Hawk, making their formation two Chinooks, three Black Hawks and two Apaches.

They lifted on time, transitioned and then the lead Black Hawk started going the wrong way. They were going in completely the opposite direction to the set plan.

Daffers said to Tweedle Dee: 'Are we following the right group?'

Tweedle Dee looked as confused as he did, so Daffers called the aircraft ahead on the radio: 'Bambi, Flipper – can you glimmer up for us?'

The aircraft turned on its lights.

'Yep, that's our bloke,' he said to Tweedle Dee.

'We are following the right people, we are just going in the wrong fucking direction. Call that Lieutenant bird in the front aircraft and find out what the fuck is going on.'

'Bambi leader, Flipper, please can you verify our directions,' he said over the radio.

'Flipper, Bambi leader,' she replied. 'We are doing the route in reverse.'

'Bambi Leader, Flipper, anyone think to share this info with us?'

'Negative, it would seem that was overlooked. Sorry, guys,' she answered.

'Thanks very much,' said Tweedle Dee. 'Shit, that's fucking brilliant. Right turn the map the other way. It did look like this and now it look likes that!' said Tweedle Dee as he tried to rejig the new route in his head so that he could give Daffers the nav. He reorganized the waypoints in reverse, which kept his head down for a while but he was quickly able to get them back on track.

At the same time they were being fully exposed to the Americans' inability to adapt to different flying conditions. The latest Black Hawk addition wasn't used to flying in formation at night. The Black Hawk was very IR suppressed on gogs. It was like following the aviation equivalent of a drunk driver. He kept slowing down. Daffers was trying to see the two aircraft in front of him and he knew they all needed to keep together.

Tweedle Dee was freaking out. 'What the fuck is he doing? Prick! Just go past him, just go past him.'

'I'd love to go past him,' Daffers said. 'You know I can't go fucking past him.'

'He's doing my fucking head in!'

Then the Black Hawk pilot called them up on the radio.

'Err, Flipper, could you turn your lights up. I seem to have lost my visual on you.'

Tweedle Dee was losing it. 'Of course you have lost sight of us, you fucking dickhead, because you are all over the fucking shop. Rubbish!'

He eventually sorted his shit out. They landed on once again at the LS. It was uneventful as the Black Hawk pilot had managed to park up much tighter. Although it was a bit of squeeze there was sufficient space to get all five cabs on the ground. They offloaded the troops. Zippy did a brilliant job. He had the hardest part to execute as he had to get alongside Daffers to get into the landing site. He really worked hard and delivered a masterful piece of flying to get his cab in and not cause an enormous helicopter pile-up.

They lifted out and returned to KAF to pick up all the guns and fire support. They were now going back in for the third time. They still didn't know that an aircraft had been shot down.

When they landed on, and after dropping the guns off, they were relieved it was over, Having been on the gogs for six hours they were hanging out of their arses. Daffers and Tweedle Dee had shut down the aircraft, unstrapped and with the helmets in their hands were standing in the cabin talking with the crewmen.

The COJHF(A) came on to the back of the aircraft and said: 'Well done, fellas, that was great stuff. I am really sorry about Flipper 75.'

Daffers looked confused. 'What's wrong with Flipper 75?'

'They are all dead,' he said.

'What?' said Daffers.

Suddenly it all clicked into place for them – the explosion, the random Chinook tailgating them and the engineers running out. Daffers felt numb. 'Shit, what happened?'

'I am sorry to have be the one to break the news to you lads but Flipper 75 was shot down by an RPG.'

While leaving the landing site, Flipper 75 was struck in the nose by a RPG that had been fired straight up at it. The aircraft nosed over and crashed. The only good thing to come out of it was that they all were killed instantly and wouldn't have known much about it.

'How many were killed?' asked Daffers, still stunned.

'All of the American crew of five and two passengers, a British army photographer and a Canadian,' replied COJHF(A).

Everybody was struggling to take it in but Daffers was really knocked sideways. He couldn't believe an aircraft had been shot down. On the last det *he* had been the RPG magnet and now at the start of this one he had just flown an operation where an RPG had actually taken down a Chinook. But we'd all known it was a case of when, not if. The Taliban had finally achieved their goal.

'Did they secure the landing site?' Daffers asked, dreading the answer.

'Yes, fortunately — because they had just offloaded the troops they were able to quickly secure the site.'

'What a relief!' said Daffers, clinging on to the news that the cab wasn't loaded with troops. 'I had visions of it being played out across Al Jazeera and the Taliban jumping for joy.'

The Americans stopped flying for a few days and the Brits and the Dutch picked up all their tasking. The whole of Kandahar felt the loss. The worst had finally happened.

Straight afterwards, Daffers and a few of the guys went over to the American crew room, the place they hung out when they were not flying.

'Hey, are the Amercian Chinook crews in there?' they asked an American senior officer who was stood outside smoking a tab. 'Do you mind if we go in?'

'Yes, I think they would really appreciate that.'

They tentatively went into the crew room where all the crews had convened. The Chinook crews and engineers were sat around deeply distressed. It was like walking into a morgue. An eerie silence hung over the room and a couple of the guys were crying quietly.

Daffers walked up to where they were all sat and said to them: 'Hey, fellas, we are a bunch of Chinook guys as well, we were on that job, we really feel for what you are going through. It sounds really hollow but if there is anything we can do, then please shout.' They nodded and he said: 'You know in Britain when someone dies we all go to the mess and get drunk on their bar bill, which is then written off. This goes on into the night and then when we are really drunk we burn a piano outside the mess. That seems to work for us.'

They didn't say much in response. A few nodded.

With that the guys excused themselves. The piano-burning tradition came from WWII days when aircrew were considered to be 'home' after crossing the piano keys at the end of the runway. If a crew didn't make it home a glass of beer was placed on top of the piano in the mess, for each member lost, along with a personal article from that person's kit locker. At the end of the wake the items would be burnt. Apparently, one time things got out of hand and the piano was burnt as well because one of the lost aircrew used to play the piano in the bar . . . and so the tradition was born and has continued ever since. It doesn't make much sense to anyone really. I certainly can't imagine what the grieving Americans made of it. I think they knew it was well meant.

I was still covering IRT at Bastion and I was pretty keen to be at KAF with the guys. We felt isolated and out of the loop because we didn't know what was going on. But at the same time there was work to be done as guys on the ground were pushing out. It wasn't long before we got our first shout, although it wasn't one of our troops that was injured.

Chapter 37

There were two rings on the field telephone. I answered it: 'IRT.'

'Sir, we have a nine-liner.'

'Gents! We have got a shout.'

Again the system spooled into action. But this time it was different. I headed over to the CP. There was no time to get an update on the downed cab. We were being called to the area of the Lastay insert in response to the All American Division, as they had an enemy casualty. We knew that he was Taliban and that he was a senior commander. They wanted to bring him back for questioning.

We raced to the aircraft and with our AH escort routed to the area. Despite all my experience of being shot at – Sangin, Musa Qala, Gulf War II – I had never felt more tense and exposed than I did on this day. It was different because a Chinook had actually been downed in the area that we were now flying to. Who was to say that we weren't next?

As we were coming in over the top of the valley, off the high ground, we could see all the locals thinning out from the Green Zone up the hill because they didn't want to get caught up in the crossfire. We could see Taliban mixing in with the civilians for protection. 'Look at that,' I said to Steve, 'the sneaky bastards.' We watched the Taliban walk up the hill in the relative safety provided by their human shields.

The Americans chucked a smoke grenade out as we were coming into land. I looked across and there were wires on the right-hand side.

Steve in the left-hand seat said over the intercom: 'I have got wires on the left.'

And then Redders on the number two piped up: 'Yep, and I have got wires on the undershoot.'

They were expecting us to land in a triangular grid of wires.

'I can't fucking land here,' I said.

The Americans were on the ground marshalling us to come down.

'Mate, it's the size of a fucking postage stamp. You can fuck off if you think I am going to land there,' I said over the intercom. 'I am coming off right. We are going to land on the other side of the wires.'

We landed on, in a furrowed straw field.

'Ramp down,' confirmed Tommo.

I looked left as the Americans brought the injured Taliban out. He had received a gunshot wound to the chest.

'Casualties coming on.'

'Has he got an escort?' I asked.

'Doesn't look like it, but we have got our guys back here,' said Tommo, referring to the five force protection guys that were with us.

'Roger that.'

The MERT got to work tending his gunshot wounds.

We lifted and routed back to Bastion. The Teletubby was going to receive the best British medical attention. I always found it weird that we shot people and if we didn't kill them it was the Brit medics that patched them up. This was the nature of war in a civilized society. The Taliban would certainly not offer our soldiers the same compassion.

I was two days longer in Bastion and then was sent back to KAF. I had been there a matter of days when Nick asked to see me to talk about the extract.

Chapter 38

I had just got back from the gym, when Redders popped his head round my door.

'Mark, the boss wants to see you in main.'

'Roger, on my way.'

I showered, threw on some rig and jumped in a wagon to JHF(A) main. Nick was in the planning room.

'Hello, mate, you wanted to see me.'

'Mark, I need you to plan and lead the extract for Lastay.'

Excellent, I thought. 'Shall we discuss it over a Red?'

We walked outside and lit up.

'Before we get to the extract, what was your take on the infill and Flipper 75?' I asked him.

'I was in the Ops room of Taskforce Corsair when it went down. You would not believe the number of feeds they have in there from the different platforms, so they knew almost instantaneously that an incident had occurred but it was impossible to know whether anyone had survived or not.

'It was only when the first wave had returned to KAF that they realized for sure that the aircraft had been shot down and that everyone onboard had been killed. TF Corsair had to reassess what they were going to do for the second and third lifts. But, as I said on the phone, the first set of troops had been inserted so there was no choice but to carry on. Otherwise it would endanger the lives of everyone already on the ground.'

Nick went on to explain that there was a limited amount of information at the time on where and how the aircraft was shot down. All the Intelligence was saying was that it was near the landing site. But as the Intelligence Officer didn't know precisely where and when it was shot down the Operations Officer had to assume a rough area for the downed aircraft. They then placed a no-fly zone around this area. This meant that on the second wave the route was reversed. This was the information that never reached Daffers, but it wasn't too bad because they had all the routing and could quickly adjust to a route reversal.

'I got to Zippy's cab,' he continued, 'and said: "Guys, I have some bad news. One of the American Chinooks has been shot down. A decision has been taken to fly the reverse route. If you stick to your tactics of going in and out as quickly as possible then you will be fine." Evs came up to me yesterday and said: "When you told me that the cab had been shot down and we were going back in, with the threat of facing death, that has to be one of the lowest points of my career."'

'You can hardly blame him for that. Do we have any more information on the threat?'

'It was non-specific in the brief. They knew that there wasn't much activity, but you know how it is, mate, it's impossible to legislate for a lone gunman or a bloke with an RPG who happens to be on a track or in a valley that is flown over. It's looking likely that the Taliban gunman got lucky. He just happened to be in the area and for whatever reason they flew over him so he was very close to them and got a clear shot.'

'You only have to be lucky once,' I said to him.

'Mark, one last thing.'

'Yes, mate.'

'There is an Army Major who is trying to file a formal complaint against you for bullying him in the toilets.'

'What the fuck. Are you serious?'

'His CO doesn't want it to reach that point and we both

hoped that if you apologized then it would all go away. Before we take it any further, I need to know – did you swear at him?'

'No, I told him to get out of the toilets because he was the second NSE blunty I had caught crapping in our traps.'

'Did you raise your hand to him?'

'No. Mate, I was hardly armed and dangerous. I was wrapped in a towel. However, if I had taken my towel off then he might have felt threatened!'

Nick smiled tightly at my poor attempt at humour. 'It's not really a laughing matter. It's more a pain in the arse that we could all do without.'

'Look,' I said, 'I am not going to apologize but I will meet with the guy and have a chat about it to see if we can get it dropped. What a big girl's blouse.' I shook my head. 'I really thought the British Army had more mettle than this!'

'Ok,' said Nick. 'I'll set it up.'

Despite the long shadow cast by the lost CH-47 we all still felt we had a job to do. I was looking forward to getting my teeth into the extract planning.

There were lessons to be learnt from the insert. In one of the first planning meetings we were all sat around the briefing table in Taskforce Corsair. I was there with two of the Cloggy 47 drivers and the Taskforce Corsair Ops Officer, a US Army Major. He was an approachable guy who was willing to listen to a different point of view.

The object of the plan was for three Brit, six American and three Dutch Chinooks to extract 500 troops from the Upper Sangin Valley. The troops had gone in, fought, expanded out, dominated the ground and then had folded back in preparation to be lifted out again.

At this initial meeting we discussed the routes in and out on the Americans' graphical multimedia mapping system. I was looking at the maps and saw the selected route. 'I don't think this is a bright idea,' I said to the Major.

'Why, what's the issue?'

'It's too obvious and therefore too dangerous.'

The two Dutch pilots were nodding in agreement. I got on well with them because we sang off the same songsheet with regard to routing and tactics.

'Right then, what are your thoughts?' he said.

'How many waves are we looking at?'

'At the moment we are looking at two waves,' he said. 'Wave one will be three Brit and three US CH-47s. Wave two will consist of four US and two Dutch CH-47s.'

'What height are you planning this at?'

'Yeah, mid-level as normal,' he said.

'We are not flying at mid-level, mate. I will tell you that now,' I asserted. Daffers' six-hour NVG ball-ache was not something any of us wanted to repeat.

'If you have to fly around at mid-level then put the Cloggies with us and we'll have the Cloggies and the Brits and we'll do our own thing. Or alternatively, you do what we do,' I said.

'Well, what do you do?' he asked.

'We come out low, climb to height and then we let down from a set distance from the target and we come in low and fast. We don't ponce around at mid-level, mid-speed. We then come out fast and low. Then we climb back up to height,' I said.

'I am not sure we can do that.'

'We agree with our British friend. These tactics are more suitable for us too,' said one of the Cloggy pilots.

We didn't labour the point and I left it with the Major to mull it over while we carried on with the planning.

The next thing we needed to look at was the landing site.

There were many questions: was it big enough? What was the surface like? Was there a slope? We needed to identify some good lead-in features and the appropriate release point.

I was shown satellite imagery of the LS and it was a series of furrowed fields. There didn't appear to be much of a slope and it was big enough to get six Chinooks in. The beauty of an extract was that the lads were on the ground providing protection for when we came in. Having said that, Flipper 75 had gone down on the way out from the drop-off point.

What had surprised everyone about this was how early in the infill it had occurred. Normally, we would expect the later aircraft to be shot. Looking at it from the Taliban perspective they think: *Ooh, hello, there's a helicopter.* And then when another one flies over, he thinks: *I know, I'll get my gun out.* And then when the third one flies over, maybe he'll have a go and shoot at it.

But in the case of Flipper 75, it was the second aircraft in the first package when they were flying in close formation so it was more unexpected. I wondered if the attack had been pre-planned rather than opportunistic. This didn't bear thinking about because that meant that the Taliban had known we were coming.

The biggest issue was going to be getting six 15-tonne helicopters into a dusty landing zone at the same time without any of them browning out. We needed to factor in the wind direction and strength and an overshoot plan for each individual aircraft in case they had to abort the landing.

Over the following days we continued to refine the plan. I was not flying and dedicated all my time to planning the extract. The biggest hurdle we had to overcome was to ensure that each wave left the release point, at the right height, at the right speed, in the right formation, giving us the best chance to affect a simultaneous six-aircraft landing at the required time on target.

Within the planning cycle we started at the pick-up and worked backwards. We decided on a landing and overshoot plan that ensured we remained safe and that all involved were happy with. We were given the time on target that the ground troops were working on. It was now a simple mathematical solution to deduce the time we needed to hit the release point and, therefore, the KAF departure time. But we still needed to address the issue of height.

Flying at mid-level did not work for us. We considered it unsafe because of the threats from the Taliban. We preferred to fly at height out of small-arms range. US Army aviation had a different take on it and I found it surprising because I knew the US Marine Corps adopted the same principles as the Brits.

The way the American SOPs worked meant that it was difficult to change them in theatre. They are by nature very prescriptive in what they do and they adhere very firmly to their rules and regulations. They tend to need a huge body of evidence and a great deal of staff work to change anything.

I had a very open discussion with the Major and explained that these tactics had been developed over many years and mostly from our experiences in Northern Ireland and other theatres.

He could see the logic but at the end of the day it wasn't his call. The aircraft that we fly and the aircraft that they fly are equipped in a different fashion. There were certain limitations on what they could and couldn't do.

But because we had been working very closely together there was a degree of trust. And the American, to his great credit, agreed to apply for a change. However, it had to go all the way back to the Pentagon to get approval, and while it eventually came through, it was authorized only for the extract. But it did enable them to come in at the same height as us, to fly a little quicker in and out, and to leave the LS faster than normal, which made the extract a lot safer.

Now that we had agreed this, the rest of the planning followed quickly, covering deconfliction, fire support, command and control, and communications. Over twenty aircraft in total would be used in the extract. In addition to Chinooks, there would be AH support and Black Hawks. One of these Black Hawks would be the Air Hammer and would carry the Colonel in command of the extraction. It was the Command and Control aircraft and it would fly at the rear of the formation.

We would have both Brit and American Apaches in support. The Brit Apaches would be on one side and the American Apaches on the other side of the delineation line. This imaginary line ran down the middle of the pick-up zone. The Brits were to come out of Bastion and the Americans were to come out of Kandahar. The route would take us to the north-west of Kandahar to the Upper Sangin. There were two routes up to two separate release points, one route and one release point for each wave.

Two days before the extract, the Brit crews attended the Air Mission Brief, the aviation brief for all participating aircrews. The Corsair briefing room was packed. We ran through the whole plan from the weather all the way down to landing back at KAF. The brief finished with the Taskforce Corsair chaplain saying a prayer, ending with his obligatory shout of 'AIRBORNE!'

'All the way, sir!' the Americans in the room responded with gusto. The Brits and the Cloggies all looked at each other in perplexed amusement. They only thing that remained for us to do was the ROC drill.

The next day Steve, Redders, Tommo and I wandered over to the large hangar that Nick English had mentioned to me in Bastion to take part in the ROC drill. As he had described, it was a surreal experience. The hangar floor was covered with rocks, tape and rope, set out to depict the terrain of the Upper Sangin Valley. The routes into the pick-up zone were marked by blue tape.

I knew this was a common practice for them because of my four years on Cobras. But it was late evening and freezing cold. It was the last thing the lads wanted to do.

'What the fuck is all this?' said Tommo.

'This is a walk-through, talk-through. Stick with it. It will pay dividends.'

The US Marine Corps called it Dirt Diving. I agreed with doing rehearsals, because walking through set the plan clearly in everyone's minds. It worked. I can understand why the Americans did it because the Colonel who was authorizing the whole mission was standing in front of the brief watching. Part of his release authority was to watch the walk-through to make sure that he was happy with the plan before he authorized the mission to go.

The American Major was running the drill. Every participating element was represented by a person who assumed a position on the map. We had to shuffle through the formation as a group of four and walk through the procedures and routing. The crewmen were less impressed and muttered through the entire experience, complaining about what a waste of their life it was. But to co-ordinate over twenty aircraft moving around the skies at one time, the pre-planning needed to be thorough, and this was the best way to ensure that it was, despite the crewmen's disdain.

Chapter 39

On the day of the extract Nick Geary was invited to the American memorial service for the five Americans, a Canadian and a Brit who had been killed when Flipper 75 was downed. It took place in one of the hangars, which was brimming with mourning American aviators and service personnel. He had given us strict orders that none of us were to attend.

As much as we respected that each country had different ways of dealing with tragedy, we felt that theatre was no place for elaborate memorial services. It seemed to us much better to deal with those emotions back at home. The grieving process gets in the way of your mindset. We needed to keep our eyes on the ball to stay in the game more than ever now. We couldn't let our feelings get in the way otherwise we might lose another cab and this time it could be a British Chinook. That would be a loss unacceptable to everyone in theatre, to Odiham, to the families, to the media and to the country. We couldn't even entertain it and yet it hung around in the back of our minds. The stakes had increased tenfold. And, while it may not have been our style, hats off to the Americans – they weren't going to let that put them off honouring their dead.

The pressed, immaculate uniforms, helmets and rifles of the deceased crew had been carefully laid out flat on the floor. The congregation gathered around them and prayed for the souls of their dead colleagues and fellow servicemen. They sang hymns and gave tribute to their lives and memories, sent their thoughts to their families and grieved at their loss.

At the end of the service they held a roll call for the squadron, shouting out loud the name of each member of the squadron present, and of those who were not. Each squadron member bellowed back: 'YES, SIR!' at the tops of their voices. When the Colour Sergeant reached the name of a deceased member he yelled out his name. The hangar echoed with silence and there was no response. A second time he shouted, even louder, the name of the dead serviceman and still the silence reverberated, ringing through the ears of the assembled mourners. Finally, for a third time, with all the energy he could muster, he yelled out the name. The silence penetrated their ears once more. This ritual was performed five times, once for each of the serving members of the squadron who had died.

Nick was in tears. The service had a colossal impact on him. It was incredibly emotional. Immediately afterwards he had to walk out of the hangar and send us off, back out into the theatre to do the extract, telling us that we would be all right. It was a difficult thing for him to have to do.

The mood had changed substantially after Flipper 75 had been lost. It put additional pressure on everything. We were always vigilant but we always had to expect that the worse could happen. As we were lifting at 2300 everyone went off for some scran.

We were joined by Nick. Everyone was looking pretty gloomy. It was very tense. The guys were nervous, mainly because we were going back into an area near where Flipper 75 was shot down. I tried to lighten the mood and break the ice by making glib comments: 'Everyone's a bit serious aren't they?'

They all forced a laugh out and stretched a tight smile. 'Ha ha.' They weren't enthusiastic.

'Fair enough,' I said and returned to my dinner.

Nick piped up. 'Look, guys, if you stick to the plan and fly like you have been trained to, you will be fine.'

Nick was great. It was good to know that he had our backs. I couldn't handle all this waiting around and feeling sorry for ourselves. It was better that we just cracked on and got on with it. But there was no getting around it. A Chinook had gone down and we were about to head back to the killing ground.

'Mate, I set up that meeting with the whingeing pongo,' said Nick. 'And I have just had a call that he's waiting for you at Pinchettos. I know the timing sucks but can you grab a Landie and go and speak to him.'

'Yeah, no worries.' I was in no mood though. It was pathetic. I opened the door into Pinchettos and he was sat in the corner. I walked over and sat down with him.

'We need to make this quick as I am lifting in a couple of hours. What's going on?'

'Look, I felt threatened,' he said.

'What do you mean, you felt threatened? I am a Bootneck. This is what Bootnecks do.'

'It was out of order. You were being imposing and threatening.'

I sat back and looked at him. *Mate*, I thought, *you have no concept. There are blokes out there having their arms and legs blown off, being shot at daily by the Taliban and you are feeling threatened by an almost naked Bootneck telling you to get out of the heads. Grow some!*

When I spoke I tried some diplomacy. 'I still believe in the point that you shouldn't be pooing in our heads. But why didn't you just come and talk to me? I think a formal complaint is a bit unnecessary. Mind you, I still would have said you shouldn't have been pooing in our heads.'

'Ok, so we agree to disagree,' he said. "Oh, by the way, we have met each other before.'

'Oh really?'

'I was at Gerry Osborne's barbecue last year.'

'What, so you know me socially, we have had a beer together and you still wouldn't come and talk to me! Mate, can we consider this dropped now.'

'Yeah, of course.'

'Good. I have to go now.'

I walked out of Pinchettos. *What a prick*, I thought, but at least that was the end of Poogate. Now all we had to do was get the troops out of Lastay without getting our arses handed to us.

Finally, we made our way to the aircraft. The darkness cloaked the scurrying activity round the cabs, but against the faint glow of the landing lights and KAF in the distance, we could see the silhouettes of the helos lined up on the dispersal. It was an impressive sight. Our trusty steeds, ready to get back to work and bring the troops home.

Once we got to the cab everyone's mood lifted because we were back in work mode. We had something to get stuck into. The one thing that we didn't want to happen was for our cab to go tits. Although we had a contingency plan, there would be a whole faff of swapping kit and reprogramming the routing, which would take Steve fucking hours.

Imagine a really good TomTom system, make it really big and very expensive, and take away most of its functionality, leaving a green screen with some numbers to tell you how far you have to go to your destination and what direction it is, and that is our nav kit. It is very good and robust, and can survive being attached to the vibrating thrashing beast that is the CH-47, but it's quite basic. Still, it gets us from A to B and any points in between if required, but we have to manually input all the data and that doesn't suit complex sorties with masses of waypoints.

Steve did a quick walk round and I signed for the aircraft. I walked up the ramp. I pulled on my LCJ, then hauled myself into the cockpit, squeezed myself into the right-hand armoured seat, put in my ceramic chest plate, zipped it up and pulled my five-point harness together, slotting the lugs into the harness box between my legs.

I moved my head from side to side, cracking it into place. I stretched out my arms, reached for my helmet and put it on my head. I lowered my NVGs over my eyes. The helmet fitted snugly on to my head. When you start flying you are fitted for your own bespoke helmet. It goes with you everywhere. It's a really personal piece of kit that is never shared, and over time it fits like your comfiest pair of shoes. It is regularly serviced and refitted to prevent the rot that hours of accommodating our sweating heads can cause.

Before firing up the beast I looked back and shouted the familiar warning: 'Helmets!' and I switched on the battery. Instantly the RADALT audio warning rang through our ears. I leant down automatically and cancelled it on the collective lever.

I glanced at the inverter, to ensure I had '6-2-3-6' on the CAP.

The middle of the centre upright console is dominated by the CAP. It contains the myriad of red, amber and green warning lights that are attached to sensors all over the aircraft. If a warning light illuminates on the CAP, depending on its seriousness flashing attention-getters alert the pilots and a klaxon goes off. This generally ensures that focus is fixed on the CAP to establish exactly what the problem is.

Sam confirmed that the APU was good to go. 'Clear PU.'

'Roger that.' And I flicked the switch. The fuel ran into the APU: *One . . . two . . . three . . . four . . . five*, then I hit 'start'.

The APU wound up and clicked over, followed by the

APU 'on' light lighting up and then the CAP light. I was relieved to hear the pitch of the APU change. So far so good.

'APU gens on – PTUs one and two,' I yelled. 'Ready panel.'

Redders pressed the maintenance panel test button at the rear and the CAP lights blinked.

'Six on,' Steve reported, then 'Six off' as he freed the switch once more.

Steve leant to his right and turned on the radios, nav kit and the cab's self-defence suite, located on the central console.

'Fire warning test.'

The klaxons deafened us.

'Loud and clear.'

'Clear the dash.'

The flight controls were moving freely so I started the number one engine. The powerful Lycoming turboshaft engine roared into life. Once the noise behind me had steadied, I started the rotors.

'Clear rotate.'

'Clear all round to rotate,' came the reply.

The blades engaged, moving slowly at first, but then faster and faster until they were turning with a regular beat.

'Clear flight on one, please.'

'Clear flight one.'

Cleanly I advanced the throttle on the number one engine to engage the power up as the blades spun into a mesmerizing rhythm of velocity and noise. I then started the number two engine.

'It fucking works! Brilliant.' A wave of relief washed over me. We could now get down to business.

As the cab was running, we all sat listening for the radio check and waiting for the timings so that we could lift in accordance with the agreed sequence.

'Five minutes to check,' I said.

We waited.

'Two minutes to check ... thirty seconds ... five, four, three, two, one.'

The radios crackled into life.

'Flipper check victor.'

Everyone in our wave responded in turn.

'Check Fox Mike.'

Again everyone responded in turn.

The plan had changed subtly. The two Dutch Chinooks were now in our wave. There were six of us in total, two Brit Chinooks, two Dutch and two American.

The radio checks completed, we taxied out on to Foxtrot.

I tugged out the master arm pin, went through the pre-take-off checks and then with my left hand pulled up on the collective. Once again I was exhilarated by the feeling that comes from pulling the power. I could feel the aircraft responding as it ground upwards. The wheels lifted and we were in the air.

As we were taxiing out, we were looking to make sure that everybody was clear. I pulled up on the collective and pushed the cyclic forward. Accelerating into the darkness, the cab throbbed with intent. Around me our loose formation of helicopters climbed and routed to the north-west. Once we were at height we ran a weapons check. The tracer was visible sequentially across the formation as each cab checked its weapons when it reached the designated point. Nothing was said over the radios. We could see the tracer coming out in shards of green light.

'Look, he is checking his guns,' said Steve.

'Really, you don't say,' was my sarcastic reply.

There was some banter in the cockpit but it was dampened by the tension that hung over us. As we retraced Flipper 75's path, her fate wasn't far from anyone's mind.

But it *was* a relief that we were flying at height, not mid-level on gogs for hours on end. On the bright side, we knew that we were landing into an LS that had been secured by the troops on the ground, which mitigated some of the risk. The ground troops were folding back towards the pick-up zone. It was possible that the enemy was shadowing their movements. The threat was still very real and the stakes very high.

We reached the release point into the descent. We were faced with a situation where six helicopters were coming in from height at speed into low level. This would be fine if we were flying in formation with five other Brits but we were not. We were flying in with Cloggies and Americans and we were all due to hit the target at exactly the same time. It was imperative that we did it all together.

We were coming down from height at speed like six rotating elephants all trying to land simultaneously on a pin head at a hundred miles per hour. It must have been a phenomenal spectacle to watch from the ground. In the air it was tense.

We had eyes on stalks.

'I have lead straight ahead. Where's two?'

'Right one o'clock,' said Redders.

'We're descending.'

'Roger that.'

'Where's my wingman?'

'He's over there at the half past three,' said Tommo.

'Can you see the Cloggies?'

'They are descending behind us,' said Tommo.

We were constantly checking and rechecking the positions of the aircraft as we all simultaneously lowered on to the LS.

I was thinking: *Check, check, check, where is everybody else in the sky? We don't want to do the enemy's job for them and bring an aircraft down because we have flown into each other.*

The period of the descent was like spinning plates. We were

looking at all the aircraft at once. Once we got to the release point at low level we levelled off. We could see everybody.

'Great, brilliant, everyone is together,' I said. It was a huge relief.

Now we had to do the tricky bit, which was land in a big dusty field with six aircraft kicking up sand. As we came into land, it was pretty fucking dusty.

We were booting in. I had looked at all the satellite pictures so I knew where the LS was. I had selected my three markers. These were the three prominent geographical points that were going to lead me into the target: a group of buildings, a prominent piece of high ground and a bend in a wadi.

'Right, I have got the buildings.' I remembered them from the brief. I could see the mountains. My landing point target was the eleven o'clock. I could see the lead. He was now starting his approach. I knew this because I could see his rotor disk angle changing, telling me he was decelerating and going into land.

In my head I was thinking: *I hope it's not too dusty. If it is then I am going to have to wave off. We have briefed the wave-off route. That means I am going to come off left. Is that the way I am going? Yep, that's the way I am going.*

All of this information was rattling around my head. Landing was the most dangerous point and the risks were heightened being surrounded by so many aircraft doing exactly the same thing. Not to mention the hundreds of troops on the ground we were about to pick up. It was stress central.

'Have you got any troops?' I asked Steve.

'Yes, I can see the troops.'

'Can you see the Cyalumes?'

'I can see the Cyalumes. Excellent. I am starting my approach. Coming down.'

'One hundred feet, thirty knots.'

'In the gate. Bugs are reset.'

'Where are the other cabs?' I asked.

'Yeah, they are good,' came the quick reply. I trusted Tommo's judgement.

'Starting my approach.'

I set the nose-up attitude and eased off the power.

Shit, I hope I don't brown out too early. Is my speed good? I am thinking.

'Fifty feet, twenty-three knots, nicely in the gate.'

'Dust cloud forming.'

'It's at the ramp.'

Redders picked up the patter: 'At the door.'

The dust cloud enveloped us.

'Aft wheels on.'

I levelled the aircraft and applied the brakes. We were down.

'Excellent job, guys, nice one.'

As we sat there turning and burning the dust cleared and we could see through the green hue of our gogs that all the cabs were on the ground around us.

'Brilliant, hoofing! Everyone has made it in one piece.'

For a few seconds I breathed a sigh of relief. But not for long.

'I want this done in no more than two fucking minutes,' I told the lads. I was sat there tapping my toes and the seconds passed rhythmically. After two minutes flew by, I was getting agitated. There were six other cabs waiting to lift out of here simultaneously and I didn't want to be the one that held the departure up. Apart from anything else we were an enormous target.

'What's going on?' I said on the intercom.

'Errr . . . we have got a problem,' said Tommo on the ramp.

'What's the problem?' I said.

'They have just driven a quad and trailer into a ditch.'

'So what the fuck are they doing?' I said getting irritated.

'Well, they are trying to drag it out,' said Tommo.

'It's TWO minutes, gents!' I said.

'Air Hammer, this is Splinter 24. We have a problem, we need more time. Request rolex plus two for lift.'

We sought permission from the airborne command and control to delay the take-off. Additionally this let everyone else in the formation know that we were delayed.

'Splinter 24, Air Hammer, roger rolex plus two.'

As the driver had approached the cab in haste to get onboard he hadn't noticed the furrowed ditch and planted the quad's front wheels in it. The quad's trailer was equipped with mortar-locating radar, which the Taliban would have loved to get their hands on. The troops on the ground were very reluctant to leave such a Gucci bit of kit as a gift to the enemy.

Thirty US paras were all gathered around trying to haul it out.

'Tommo, mate. We have to fucking go. You are just going to have to tell 'em,' I said. 'There are six aircraft waiting for us, it's fucking stupid!'

'I know . . . I know . . .' he said frustrated.

'FUCKING LEAVE IT, C'MON, WE HAVE TO GO! C'MON, C'MON, C'MON!'

He was yelling at them at the top of his voice. In the end he legged it off and physically grabbed them and pushed them on to the ramp.

'WE HAVE TO FUCKING GO! GET ON THE FUCKING CAB!' Tensions were high.

We couldn't hang around any longer. Eventually we got them all aboard.

We lifted off. As we came out of the dust clouds, I was constantly checking the area.

'Where's lead . . . where's two, where's three, where's four? Have we got everybody?'

'Yes, we have,' said Redders.

It was like trying to herd six kids to the toilet in a packed supermarket. As we all safely reached height it was a huge relief to have got the trickiest and riskiest bit done and dusted. The formation fanned out as we ascended. We now had the space and freedom of the heavens to stretch out in.

As part of the plan, Simmo was to fly a spare CH-47 in formation with the Air Hammer, which was holding to the east, flying around in silly circles. Simmo was working hard to keep visual contact with the smaller, less visible aircraft.

There was a lot of radio traffic, discussing what to do about the stuck quad and trailer, and Simmo called me: 'Splinter 24, Splinter 26, Air Hammer want me to go and pick up the quad and trailer. Confirm its grid,' he said.

'Splinter 26, this is Splinter 24, that's ludicrous. You can't go in and get it. Let me speak to them. Stand by.'

I got on a different radio.

'Air Hammer, this is Splinter 24. I advise strongly against using the spare. The quad bike and trailer are in a ditch and could not be moved by thirty guys on the ground. If thirty troops couldn't pull it out of the ditch then one or two crewmen are definitely not going to be able to move it.'

'Roger, that,' came the reply. Common sense had prevailed.

The decision was taken that the quad bike and trailer would be blown up in situ. There were two Brit Apaches to the west of the LS and two American Apaches to the east. Suddenly it was like *The Cannonball Run* between the four Apaches to see who was going to get permission from Air Hammer to fire a Hellfire at the trenched kit.

Nick English got to the scene first: 'Air Hammer, Wildman, visual on the target. Request clear hot.'

'Wildman, Air Hammer, clear hot.'

He unleashed a Hellfire missile and the quad bike and trailer disintegrated into a ball of fire.

We routed back. We flew south towards the Red Desert – we were going the long way round as a deception tactic. The job was done and we had successfully pulled it off without a hitch.

We landed on at KAF. Nick was waiting for us. He wanted to personally greet each of the crews as they came back. We shut the aircraft down and unstrapped.

I walked down through the cabin.

'Good job, gents,' I said to the crew.

Nick walked up the ramp.

'Good job, mate.'

'Yeah, thanks. I can't believe they deep-dusted that quad. Did you hear they had to blow it up?'

'Yeah, I heard. Let's speak about that in the morning. You guys need to get your heads down. I am replacing Steve in your crew and we are going forward to Bastion tomorrow to hold IRT,' he said.

And so the war continued.

Epilogue

30 October 2006

I knew I was finally coming home. Walking on to a C17 was like walking into a vast, polished-silver, curved hangar with a latticework of pipes. I never failed to feel impressed. The great thing about flying pax in a C17 was that it was truly a first-class experience. It was the only proper flat bed that fitted my 6ft 3in frame.

I rolled out my sleeping bag and settled in to get my head down until we got to Turkey. When we arrived in Turkey, where we had a two-hour layover, we piled off the aircraft and were bussed straight to the American store on the airbase.

'Right, here's the plan. We drink as much beer as we can in an hour!' I said.

We ran in and each bought a six-pack. We sat outside and yammed the six beers. The beer was gassy and cold. We hadn't had a drink for eight weeks and it went straight to our heads. It was great just to be torpedoing it down. Everything was slightly manic because we trying to kill time but at the same time we could not get too excited about getting home. In the military, we were never home until we were home. There always has to be an error margin as things inevitably go wrong. By the time we had sunk all six of the beers we were hoyed! We clambered back on the bus. We had to play it fairly carefully. A C17 was not a scheduled passenger plane and they wouldn't take kindly to inebriated servicemen aboard the aircraft.

We boarded, our heads down and our mouths shut so as not to draw attention to our giddiness. I climbed into my sleeping bag and as the aircraft ascended into the climb, the vibrating floor lulled me into a mildly drunken sleep. I was awoken by the civvy bloke who was lying next to me.

'All right, mate,' he said. 'You can let me go now.'

I had rolled over in my sleeping bag, snuggled in and started cuddling him. He was probably quite surprised to be amorously embraced by a tipsy, snoring, bald-headed 6ft 3in Royal Marine.

'Yeah, I am really sorry.' I unfurled myself from him, a little embarrassed, rolled over and went back to sleep.

I was awoken from a sound sleep by the guy next to me.

'Mate, we are starting our approach soon.'

I sleepily gathered up all my stuff. My mouth tasted as dry as Gandhi's flip-flops. I sat back down in my seat, along the side of the airframe, and belted up. It dawned on me that I would soon be landing. I was excited, tired and desperate to get home. In my head I was imagining the descent over the green fields of Oxfordshire, a welcome change from the barren, desert wastelands of Afghanistan.

I had spoken to the DCOS from Afghanistan and told her the joyous news that I had managed to snag myself an early flight out.

'Should be with you on Monday,' I told her.

'What date's that?'

'I don't know, sweetheart. I am on the phone in the middle of the fucking desert. Look at the calendar,' I cheekily responded.

'The 30th. No way! I can't believe it!'

'What?'

'We are not here. We are in the States! We fly out on the

29th. I decided to take the girls home for a week at half-term as you weren't due back.'

Fucking hoofing! I thought.

I arrived back to my empty house, with only Flash, my German shepherd, there to greet me. (The neighbours had been looking after him.) He was over the moon to see me, furiously wagging his tail and running around in circles. It was good to be home, even if my family weren't there.

I wasted no time and opened myself a beer. Beer in hand I browsed the net to see if I could find myself a reasonably priced flight to get me over the pond and join up with my girls. I couldn't wait to see them. It was torture to be around all of their things and not be able to hug them or hear them as they bounded around the house. The house had an eerie silence about it, which made me miss them even more.

I managed to square away a flight at early o'clock to Minnesota, my wife's home state, via Minneapolis. I threw a few things into a hand luggage rucksack and blagged a lift to Heathrow from one of my mates, Nethers. He's a good bloke and single so was willing to get up at the crack of sparrow to get me to the airport.

It's harder for the single guys. The transition from theatre to home can be very stark. These contrasts can mess with their heads. Sam summed it up nicely. 'One night I am in a cab,' he told me, 'watching a soldier having his heart massaged by the MERT, with his chest cracked open, and twenty-four hours later I am sat on my own in my living room watching *Eastenders* and David Beckham's new haircut is front-page news!' At least when I get home I am in the thick of family life. There isn't usually much time for reflection because there is always more shit to do than I can manage.

The DCOS was waiting for me at the airport and we had

one of those corny airport terminal arrival moments. I was so pleased to see her. She ran up to me and I hugged her tightly, spinning her round. Marriage is no easy business and certainly not a military marriage but there are moments when we are reunited, I forget what a pain in the arse she is and she forgets what a nightmare I am to live with. We are in love again, like newlyweds.

We chatted frivolously about the journey, how excited the girls were. It was trivial but great to be among it all again. We drove straight to the liquor store and piled up the car with booze, then drove back to Mel's house, one of her school chums, where we were all staying. I couldn't wait to see my girls. It was so frenetic. We pulled up the drive and they came running out to see me. As I opened the door and stepped out of the car, they launched at me and covered me in kisses and hugs. It was awesome to be enveloped in my family and have them coated all over me. I had missed them so much. We went inside and started drinking and catching up. The home-coming party had begun.

We didn't talk much about Afghanistan. I alluded to the fact that it had been a little sportier then I had anticipated but not much more. The DCOS didn't want to hear it. This wasn't the time and I didn't want to tell her and spoil the moments that we were enjoying. We were staying with family and friends so none of the normal readjustments had occurred. This utopia lasted a week. The cheapest flight that I could get back was twenty-four hours later than their flight. The hardest part was that I had to say goodbye to them for another day. I drove them to the airport, hugged and kissed them goodbye, then watched them all walk away through departures. I was full of sadness. We had all just got used to being around each other again and then we were separated.

I arrived back to utter madness. I had literally dumped all

of my kit on the kitchen floor and flown to America. The jet-lagged DCOS had arrived back, with all of her and the kids' luggage, to all of my kit as well. The kids had to be ready for school, they were jet-lagged, I was jet-lagged, the DCOS was jet-lagged, the dog was running around with uncontrollable excitement, tripping everybody up, the girls were tired and grumpy. Inevitably it all blew up into one emotional nuclear explosion. It was like a much-needed thunderstorm that cleared the air. It enabled us to recalibrate and settle back into the routines of normal living. The honeymoon was well and truly over and I was now home.

Soon enough I was summoned to the boss's office. Not usually a good sign.

'Boss, you wanted to see me?'

'Hello, Mark, come in, sit down. You and your wife have been invited to the Chief of the Defence Staff's drinks.'

'Have we?'

'Yes, it's up in town. Best bib and tucker, take your Lovats. You'll be put up in a posh hotel in London.'

He then proceeded to give me the timings, dress code for the DCOS and details of the events.

On the day of the drinks, the DCOS and I jumped on the train up to London, made our way to the hotel and checked in. The DCOS and I enjoyed a drink in the bar and then headed off to the reception room. We all sat down on chairs that had been laid out. A naval Captain in civvies walked out to the front of the room and addressed us:

'Welcome, everybody, I would like to thank you all for coming here today. You have been brought here under the guise of an invitation to Chief of the Defence Staff drinks when in reality it is for a different reason. Tomorrow, there will be an announcement by the Ministry of Defence about the honours and awards achieved in Afghanistan. Each one of you has been selected to represent your service with regard to this announcement. Each of you has been awarded a medal for gallantry.'

There was a pause as we absorbed the information.

'Next door are the families of two servicemen who died serving their country and have been awarded medals of gallantry posthumously . . .'

These medals were the Victoria Cross, the highest military medal for valour in the face of the enemy, and the George Cross. Their families were next door and felt that their presence would dampen the celebration of our decorations. Following this he read out the citations for each of us so that we were aware of what award we were receiving.

As the Royal Marines are part of the Royal Navy, which is the senior service, my citation was read out first. It read along the lines of:

Major Mark Hammond was an exchange Chinook pilot with 18(B) Sqn and was involved in three separate CASEVAC engagements in one night in Sep 06 in which he showed leadership, superior flying skills and inspirational command of his crew – each time under fire. The first was the extraction of a seriously wounded soldier from Sangin under fire. While Apaches provided suppressing fire he made an aggressive quick approach to the landing site and successfully collected the casualty. On arrival back at Camp Bastion he received a second call to extract a critical casualty.

Despite knowing the casualty location was under attack from the Taliban, he landed using night vision goggles while being engaged by enemy fire from several positions and the approach had to be aborted. A nearby Apache crew witnessed two rocket propelled grenades (RPGs) pass just 10 metres above and below the Chinook. Back at base four rounds were found to have hit the aircraft, one causing almost catastrophic damage to a wing blade root. So Hammond took another Chinook and, despite further sustained fire, managed to extract the badly injured soldier. For this he is awarded the DFC.

The DCOS looked at me and said: 'You did what?'

I looked at her and shrugged my shoulders. The Captain then went on to read the citations of the other awardees. It was a truly humbling experience. When they read out Matt Carter's citation for his Military Cross, I thought: *Jesus that's unbelievable!* Operating as Forward Air Controller, he'd jumped off the back of a Chinook at 20ft to get into the field and stood on top of a building in full view of the enemy in order to see the target and call the jets in so they could drop bombs on it.

There was another guy who had received the Conspicuous Gallantry Cross for running between two armoured vehicles while snipers shot at him to pick up a wounded soldier, throw him over his shoulder and run up a hill. *I didn't do anything, not compared to these guys*, I thought.

By comparison with the adversity that they had endured my citation seemed frivolous and unworthy. I had done my job and I truly believe any aircraft captain from the Chinook force would have done the same on that night. I couldn't have done it without Daffers, Sam Spence and Dan. I had written Sam and Dan up for an award. Dan was awarded a Mention in Despatches and Sam was awarded a CJO (Commander Joint Operations) commendation. I felt he should have got an MID like Dan. It was hard for Sam. He had been to hell and back. He had felt the heat of missiles as they flew past his face. It wasn't fair. We would not, could not, have made that LS without Daffers and yet he received nothing. It was strange that they were not here with me to share in this moment. I would accept this medal on their behalf. It was a team effort and it would have been impossible without them. It was a night of celebration and elation.

The DCOS and I arrived in a taxi, which dropped us off in front of Buckingham Palace. There were tourists swarming around the gates, waiting for the changing of the guard. We jostled our way through them to the main gate, showed our invitations and the security guard let us in. We were ushered into the inner courtyard of the palace. Once inside the guests were directed to their seats in the hall where we were to receive our awards and those that were to be inaugurated were told to wait further instructions.

Before we entered the hall, we were given the protocols and etiquette brief. 'Bow your head, step forward, Her Majesty will shake your hand, have a quick chat and then you step back. It will be very clear when your time is over.'

We all waited for our turn and then were directed to the Investiture Hall, a huge palatial ballroom of a splendour and magnitude that I had never witnessed before. I was humbled and in awe of my surroundings. The vastness of the room easily swallowed up the audience of family and friends, over a hundred people. As I entered the hall I was struck by the hugeness of the occasion. It took my breath away and I felt small and insignificant.

I stood next to a chap called Hugo, who I had met at the announcement evening. I was incredibly nervous. I was aware that I was about to meet the Queen and I didn't want to get it wrong. I marched out into the hall and stopped by the naval Captain as instructed. When it was my turn he gave me the nod, I marched out, stopped, bowed my head and then I was stood in front of Her Majesty the Queen. This was a moment in my life that I had never expected, that I would be stood in front of the Sovereign being decorated for gallantry.

'I understand you were flying in Afghanistan.'

'Yes, ma'am.'

'I understand you were flying mostly at night. How did you see?'

'We wear night vision goggles, ma'am.'

'It must have been very scary.'

'It was a bit at the time, ma'am.'

Her Majesty then stepped forward and pinned the Distinguished Flying Cross to my chest. I struggled to comprehend the moment. I was stood six inches away from the Queen. It was truly incredible.

'What are you doing next?'

'I go back to Afghanistan in two days ma'am.'

'Well off you go then.'

With a positive handshake I knew that my time was over. I stepped back, bowed, turned right. And that was that and she moved on to the next one.

I marched outside to be photographed. While I was waiting, I saw a couple of guys that I knew and joined them. A pilot from Odiham called Christoper 'Has' Hasler, who had also received a DFC, and an old Bootneck mentor, Colonel Bill Dunham, a stockier, older version of me, were stood around trying to see if they could spot anyone famous. We were enjoying the moment. It's not that we didn't take the ceremony seriously, it was just that we didn't take it *too* seriously. Perhaps it's just that experience in the battlefield gives you a different perspective. Or maybe we have just never grown up. Yeah, that'll be it.

Hugh Laurie was stood directly behind us. 'Has' had his back to him. He said: 'Is there anyone famous here?'

'What, you mean apart from Hugh Laurie stood behind you,' I said. And he turned round and jumped back in surprise.

Hugh Laurie came over and stopped for a quick chat. I couldn't resist. 'My wife is so happy that you are here,' I gushed. 'She loves *ER*.' I thought that was quite funny. House, MD, looked at me in utter bewilderment and walked off.

He just wanted a decent book to read ...

Not too much to ask, is it? It was in 1935 when Allen Lane, Managing Director of Bodley Head Publishers, stood on a platform at Exeter railway station looking for something good to read on his journey back to London. His choice was limited to popular magazines and poor-quality paperbacks – the same choice faced every day by the vast majority of readers, few of whom could afford hardbacks. Lane's disappointment and subsequent anger at the range of books generally available led him to found a company – and change the world.

'We believed in the existence in this country of a vast reading public for intelligent books at a low price, and staked everything on it'
Sir Allen Lane, 1902–1970, founder of Penguin Books

The quality paperback had arrived – and not just in bookshops. Lane was adamant that his Penguins should appear in chain stores and tobacconists, and should cost no more than a packet of cigarettes.

Reading habits (and cigarette prices) have changed since 1935, but Penguin still believes in publishing the best books for everybody to enjoy. We still believe that good design costs no more than bad design, and we still believe that quality books published passionately and responsibly make the world a better place.

So wherever you see the little bird – whether it's on a piece of prize-winning literary fiction or a celebrity autobiography, political tour de force or historical masterpiece, a serial-killer thriller, reference book, world classic or a piece of pure escapism – you can bet that it represents the very best that the genre has to offer.

Whatever you like to read – trust Penguin.